Agency and social structure in Human Factors

Agency and social structure in Human Factors

A critical realist approach

Roland Gasser

This book is based on a PhD thesis available online through the digital repository of the Technical University of Berlin:

http://opus4.kobv.de/opus4-tuberlin/home

© Roland Gasser 2014

Published by Lulu.com, Raleigh, N.C. USA

ISBN 978-1-312-51463-8

Contents

Tables	iv
Figures	v
Abbreviations	vi
Acknowledgements	vii

1 Introduction   **1**
    1.1   Social construction of technology   3
    1.2   Naturalistic and critical approaches in Psychology   5
    1.3   Human Factors in relation to other disciplines   7
    1.4   Contents of thesis   10

2 Computer-supported planning and scheduling in manufacturing   **13**
    2.1   Industrial production planning and scheduling   14
    2.2   The planning and scheduling task   22
       2.2.1   Planning hierarchy   23
       2.2.2   Temporal cycles and recurrent tasks   24
       2.2.3   Task analysis of planning and scheduling   25
    2.3   Enterprise resource planning systems   29
    2.4   Interdisciplinary challenges: Planning in socio-technical contexts   33

3 Human Factors in planning and scheduling   **37**
    3.1   Psychology of planning   40
       3.1.1   Planning and scheduling as decision making   43
       3.1.2   Planning and scheduling as collaborative activity   51
    3.2   Cognitive engineering of enterprise resource planning systems   54
    3.3   Some observations and intermediate conclusions   60

4 A critical theory/methods package for Human Factors   **66**
    4.1   Critical theoretical perspectives in technology-related sciences   68
    4.2   Contributions to the development of a critical perspective in Human Factors   70
       4.2.1   Technology as 'ecological niche'   73
       4.2.2   Organizational change and technology   74

	4.3	Technology in the light of Critical Thought	77
		4.3.1 Critical perspectives in Organization Sciences and Information Systems Research	78
		4.3.2 Critical Psychology, Activity Theory and Workplace Studies	82
	4.4	Philosophical re-orientation	85
		4.4.1 Critical Realism, organizations and technology	92
		4.4.2 Cognitive Work Analysis: A realist framework?	93
		4.4.3 Implications for a critical approach in Human Factors	103
	4.5	Development of an analytical approach	105
		4.5.1 Development of a regional ontology for Human Factors	106
		4.5.2 Socio-cognitive discourse analysis	109
		4.5.3 Morphogenetic approach to organization and technology	115
	4.6	Discussion of proposed theory/methods package	119

5 Study one — 120

5.1	Access and data collection	121
5.2	Analysis of planning and scheduling structures and functions	126
	5.2.1 Business unit characteristics	126
	5.2.2 Orders and planning workflow	126
	5.2.3 Problem areas and potential conflicts	129
5.3	Socio-cognitive discourse analysis and social modeling with i^*	130
	5.3.1 Description of analytical method	132
	5.3.2 Management: ERP as an instrument for flexible manufacturing	138
	5.3.3 Central production planning: ERP as an administrative tool	139
	5.3.4 Shop floor scheduling: ERP as a reality-ignoring system of justification	141
	5.3.5 Distributed intentionality and strategic dependencies	142
5.4	Intermediate discussion	147
	5.4.1 Social discourse analysis, ontology and retroduction	148
	5.4.2 Retrodiction, model building and critique	149

		5.4.3	Lessons learned	154

6 Study two — **157**
- 6.1 Approach and methods — 157
  - 6.1.1 Research hypotheses and topics of interest — 157
  - 6.1.2 General objectives and intended outcomes — 158
  - 6.1.3 Methods and research plan — 159
- 6.2 Data collection — 164
- 6.3 Results of (extended) Cognitive Work Analysis — 165
  - 6.3.1 Work domain — 166
  - 6.3.2 Operational uncertainties — 167
  - 6.3.3 Control capacity — 172
- 6.4 Morphogenetic analysis — 172
  - 6.4.1 Questions to ask — 172
  - 6.4.2 Structural conditioning / inscription — 173
  - 6.4.3 Social interaction / agential reflexivity — 176
  - 6.4.4 Structural / sociocultural elaboration — 179
- 6.5 Intermediate discussion — 181

7 Overall discussion and conclusions — **185**
- 7.1 Social structure in Human Factors — 186
- 7.2 Agency in Human Factors — 189
- 7.3 Critical realist theory/methods package — 197

8 Implications and outlook — **205**
- 8.1 Theoretical implications — 205
- 8.2 Practical implications for research and development — 208
- 8.3 Implications for teaching and training — 211
- 8.4 Possible future directions — 213

References — **218**

Tables

Table 1: Characteristics of two important aspects of Human Factors as a discipline — 9

Table 2: Overview of structural and functional interconnections in planning and scheduling (McKay & Wiers, 2006) — 27

Table 3: Comparison of planning aspects (adapted from van Wezel & Jorna, 2001, p. 282) — 42

Table 4: Comparison of context/situation and task features between classical decision making (CDM), naturalistic decision making (NDM), and production planning and scheduling (PPS) environments — 46

Table 5: Overlapping domains of the real (Bhaskar, 1975, p. 56) — 89

Table 6: Functionality and intentionality of a city (Rasmussen et al., 1994, p. 43) — 95

Table 7: Interviews and observations in case study one — 123

Table 8: Observation interview guideline for persons with planning tasks (Gasser et al., 2008) — 124

Table 9: Swim lane diagram for order workflow in business unit — 128

Table 10: Codes used to establish topical superstructure — 133

Table 11: Codes for linguistic markers of discourse (cf. van Dijk, 1995, pp. 22-32) — 133

Table 12: Codes for transformations of social practice (van Leeuwen, 2008a) — 136

Table 13: Coding statistics according to code groups — 137

Table 14: Coding statistics according to work area and code group (not occuring codes omitted) — 138

Table 15: Strategic dependencies distilled from discursive elements — 143

Table 17: Extended and modified Cognitive Work Analysis method for control tasks — 160

Table 18: Interviews and observations in case study two — 164

Table 19: Resulting control capacity overview — 171

Table 20: Variables influencing coordination of job orders — 178

Table 21: Characteristics of conventional Human Factors approaches and the proposed critical theory/methods package — 203

Table 22: Spatiality and temporal spanning of ERP studies (adapted from Koch, 2005, p. 53) — 216

Figures

Figure 1: Characteristics of generic industrial transformation process (adapted from Dessouky et al., 1995, p. 453)   16

Figure 2: Generic framework of production control (adapted from Kontogiannis, 2010b)   18

Figure 3: Structural and operational control behavior (adapted from Wäfler et al., 2008)   20

Figure 4: Goal-oriented task analysis of supervisory production control in a flexible manufacturing environment (adapted from Usher & Kaber, 2000, p. 440)   28

Figure 5: MRP-II framework (adapted from van Wezel & Jorna, 2001) 31

Figure 6: Schematic representation of collaborative production planning   53

Figure 7: Structurational Model of Technology (adapted from Orlikowski, 1992, p. 415)   79

Figure 8: The Transformational Model of the Society/Person Connection (adapted from Bhaskar, 1979)   87

Figure 9: A discursive framework for organizational change (adapted from Jian, 2011, p. 47)   113

Figure 10: The morphogenesis of structure (adapted from Archer, 1995, p. 193)   116

Figure 11: Organizational routines are generative systems (adapted from Pentland & Feldman, 2005, p. 795)   117

Figure 12: Overview of analytic steps in study one   132

Figure 13: Strategic dependency model for production planning and scheduling   147

Figure 15: Model of drift and control in ERP implementation (adapted from Nandhakumar et al., 2005, p. 238)   151

Abbreviations

AI	Artificial Intelligence
ANT	Actor-Network-Theory
APS	Advanced Planning System
CAD	Computer Aided Design
CDM	Classical Decision Making
CR	Critical Realism
CWA	Cognitive Work Analysis
ERP	Enterprise Resource Planning
ES	Enterprise System
HCI	Human-Computer Interaction
HF	Human Factors
HRO	High Reliability Organizations
IT	Information Technology
ICT	Information and Communication Technology
IOR	Integral Organizational Renewal
MIS	Management Information System
MRP(-I)	Material Requirements Planning
MRP-II	Manufacturing Resources Planning
NCR	Non-Conformance Report
NDM	Naturalistic Decision Making
OR	Operations Research
PPS	Production Planning and Scheduling
SCOT	Social Construction of Technology
STD	Socio-Technical Design
STS	Science and Technology Studies
Y2K	Year 2000

Acknowledgements

The endeavor of writing this book has been long and intense, and not always straight forward due to manifold obstacles that characterize applied sciences research within and about organizations and their technologies.

I started out exploring the persuasiveness of technologies in the field of health care around 2004 when I was a project manager at MEDGATE, a provider of telemedical services. Within this company, research was encouraged through a strong commitment to the integration of both science and practice. Our participation in the EU-funded MyHeart project, which involved 33 partners led by Philips, a consumer electronics manufacturer, was of particular importance for my motivation to become a scientist. This is where the first ideas around a PhD thesis started to form. I am very grateful to my superiors, teachers, friends and colleagues who supported me during this first phase of my journey: Anthony Dyson, Serge Reichlin, Daniel Müller, Ulrike Ehlert, Dominique Brodbeck, Markus Degen and Jürg Luthiger.

The thesis then became more focused and concrete during my time as a research assistant at the University of Applied Sciences Northwestern Switzerland. During this second phase I enjoyed the stimulating and enriching academic environment at the Humans in Complex Systems Institute. I am highly indebted to my colleagues at the institute and also within the wider environment of it, at the Technical University of Berlin and within the EU-COST funded international HOPS network of researchers (Human and Organisational Factors in Industrial Planning and Scheduling, A29): Toni Wäfler, Katrin Fischer, Helmut Jungermann, Rüdiger von der Weth, Johan Karltun, Dina Schachenmann, Simon Krauer, Dieter Fischer, Jessica Bruch, Kathrin Gärtner, Peter Higgins, Frank Ritz, Brigitte Liebig and Josef Stalder. I am also greatly indebted to the participants in our field research during this phase, and the supporters within the management of both companies that we were able to work with. It was a great pleasure to collaborate with these professionals at their workplaces.

When I successfully applied for a research fellowship awarded by the Swiss Science Foundation, the endeavor entered its third phase.

This time I was enjoying the privilege of being a visiting scholar at the Cognitive Engineering Lab of the University of Toronto. This experience led to major improvements and additions to the thesis, and I would like to thank those friends and colleagues who supported me and my work with ideas, suggestions, and discussions during this time: Greg Jamieson, Antony Hilliard, Craig Shepherd, Lysanne Lessard and Eric Yu.

Furthermore I would like to thank the Swiss Innovation Agency, the Swiss State Secretariat for Research and Education and the Swiss National Science Foundation for financial support of substantial parts of this dissertation. I am also thankful to the University of Applied Sciences Northwestern Switzerland and the administration of the Canton of Bern, my current employer, who have provided me with the opportunity to work on a flexible part-time schedule over more than four years.

The thesis would not have been possible without the generous and tolerant backing of my supervisor, Dietrich Manzey, who has supported my project throughout all these years and given timely and useful advice whenever necessary. Our meetings have taken place in New York, San Francisco and of course in Berlin – in perfect correspondence with the trans-atlantic scientific origins of my work.

My family deserves special mentioning in this place. Their moral as well as practical support and the emotional fundament they have provided me with in difficult times has been of crucial importance for my work. My parents as well as my in-laws have – among other things – helped us to organize and facilitate our two-year research stay in Toronto, for which I am highly grateful to them. Without having my loved ones around me all these years, the endeavor of writing a PhD thesis would have been a much more arduous affair. Through many discussions with my partner Sophie Vögele I have found a way of dealing with general as well as specific academic questions and doubts. Many thanks to her and also to our two amazing daughters for helping to keep my feet on the ground.

1 Introduction

> (...) the complexities of the modern workplace are such that there is a need for increased cooperation across the organizational and human factors traditions. As we have seen, nowhere is this need more apparent than in the domain of work design and cognitive ergonomics (Hodgkinson & Healey, 2008, p. 403).

Human Factors is concerned with the design of work environments as socio-technical contexts to improve their overall performance. Throughout the past 70 years, researchers and practitioners have made many socio-technical systems safer, more reliable and more efficient. The discipline has evolved into an interdisciplinary science with many applications in various industries. However, it is in a challenging situation today. Due to the increased use of information technologies, many work contexts have changed fundamentally during the last couple of decades:

- Technological systems are becoming *softer* in the sense that they consist of less hardware and more software, involve less standard operating procedures and more knowledge work, or even embrace specific culturally embedded social practices or routines (Howard-Grenville, 2005; Kellogg, Orlikowski & Yates, 2006).
- Various forms of *communication* play an ever more important role in collaborative virtual environments, distributed over increasing geographical distances, more or less structured and often within informal social networks rather than formal teams (Funken & Schulz-Schaeffer, 2008; Knorr Cetina, 2009; Nardi, Whittaker & Schwarz, 2001).
- The relationship between automated functions and human supervisory control becomes increasingly blurred in highly *networked* and predominantly *intentional* work domains (Bisantz & Ockerman, 2002; Kontogiannis, 2010b).
- Technology is developing into supporting *distributed activities* in a hybrid context of human and non-human agents, producing 'flows of information' rather than material flows (Kellogg et al., 2006; Rammert, 2007).

- Increased potential for feedback and mutual interference creates environments with higher degrees of *dynamic complexity* (Hoc, 2000; Sterman & Sweeney, 2005).

In sum, large-scale information technology systems that are widely distributed, dynamic and adaptive pose some quite fundamental issues to Human Factors. For instance, these developments are putting in question the conventional socio-technical approach of engineering psychologists and cognitive engineers who used work on problems of automation in much more straight-forward configurations (usually single worker, single machine). It becomes increasingly difficult to define a work domain, work system, or even an individual's mandate or task. In addition to that, many software-based human-machine interfaces are in a constant evolution and are highly customizable by their users.

These challenges to our discipline need to be addressed, if Human Factors is meant to evolve and adapt to new technological environments. To achieve that goal, it becomes a necessity to develop a more flexible theoretical framework and methodology that puts emphasis on human *agency* in complex work environments, rather than the search for of universal (but frequently too abstract) models of human behavior, especially within socio-technical systems that are predominantly *intentional*. If Human Factors fails to come up with such a framework, it would have to constrain itself to research and theory building around problems that are located within clearly defined boundaries, e.g. process control or similar predominantly *causal* work domains - using Rasmussen and colleagues' categorization of work domains (Rasmussen, 1986a; Vicente, 1999; Wickens & Hollands, 2000).

The purpose of this thesis is to explore limitations of current Human Factors approaches with regard to complex, distributed computer-mediated work environments and to suggest theoretical and methodological extensions to address these issues. To achieve that, it is exploring the potential of emancipatory or *critical* theoretical perspectives for this application-oriented discipline. By the term 'critical', I mean all kinds of scholarly thought that take into account the social and historical context of scientific work in a (self-)critical, dialectic approach. Without such considerations, Human Factors risks being limited to a very narrow field of use, i.e. small and relatively closed 'man-machine ecosystems' like operator contexts in driving, flying, or process control. And even these 'classical' fields of application for Human Factors knowledge are in a state of transition towards more networked, more distributed and hybrid work environments, for example in the aviation or railway industry.

The existing theoretical repertoire of Human Factors is ill prepared for the description and prediction of human activities in such complex environments (cf. Hodgkinson & Healey, 2008). Its theories apply more easily to well-structured work environments like process control, computer-aided design, or train driving. Mostly due to the restrictions imposed by the discipline, those theories are refined and elaborated through controllable experiments, where contextual influences are minimized and strictly controlled. The rigorous control of confounding variables leads inherently to an more artificial work situation within the laboratory, which in turn is seriously limiting the applicability of the results in a 'real world' setting (Valsiner & van der Veer, 2000). In addition, there is not much integration among different streams of thought within neighboring disciplines, for example between Organizational Science and Human Factors, but also with others such as Sociology of Technology or Information Sciences.

The key to overcoming these difficulties is, in my opinion, to address theoretical issues of *agency* and *social structure* in the context of technologically intense work environments. Only after having laid out the theoretical premises and assumptions, we will be in a stronger position to approach the newly emerging systems of highly distributed and networked collaboration from a Human Factors perspective. In the following, I describe some possible starting points for this project. Besides an introduction to social constructivist thought on technology I will argue for the necessity of a new Human Factors perspective that is complementary to the existing basic research and design-oriented approaches.

1.1 Social construction of technology

In order to reach across disciplinary boundaries, it is of crucial importance to discuss basic concepts. Fundamentally, our understanding of technology as a concept will influence our research and our dialogue with other scholars as well as groups involved in the design and usage of it. In the past three decades, there has been a substantial shift in social sciences concerning theories of technology.

Social constructivists have convincingly argued that technology is socially shaped and therefore cannot be understood without analytically involving the historical and cultural context as well (Bijker, Hughes & Pinch, 1987; Dahlbom & Mathiassen, 1993; Suchman, Blomberg, Orr & Trigg, 1999). As such, technology does not have an essence, it is always embedded in a social context that constitutes its use and functionality. Any scientific approach with an essentialist conception of technology therefore has to be critically evaluated and reflected, especially with regard to the fundamentally different roles of designers and users. Scientific and engineering models resulting from purely essentialist work are limited to specific aspects of a technology, and hereby ignoring any social influences and consequences. In certain environments or contexts such reductionist models of technology may not only be misleading but dangerous, especially since they do not allow for democratic control in their seemingly objective restriction to the alleged essence of technology (Feenberg, 1999; Law, 1991).

In many technology-oriented sciences this call has been heard. But, instead of rethinking the theoretical fundaments of the discipline(s), many scholars have responded by describing a variety of 'mental models' or 'implicit theories' involved in the development of a technology, hereby conserving an essentialist perspective (Bisantz & Ockerman, 2002; Vicente, 1999). For example, they have been working on the model of the engineer as developer of a technology, the model of a process or processes embedded within that technology, and the model of the user and the users' understanding of the technology. When applied to a certain case, these conceptual representations are leading to certainly useful insights, but they are still ignoring any social consequences or meaningful historical contexts. Therefore any such 'model' - especially those guiding technology development and implementation - has to be critically reflected in terms of the conditions of its theoretical fundaments, the historic and social context of its development and its critical (or uncritical) validation within a scientific or practical community (e.g. Flach, 1995). And, increasingly under the trend towards more complex working conditions described above, it becomes more important for application-oriented research to explicitly and consciously work on the basic assumptions and restrictions of its approaches, concepts and methods. This is especially the case for Human Factors as it is confronted with today's networked information technologies put to work.

Consequently, psychological or sociological studies of human activities also have to take into account the *positioning* of the observer and the observed to account for a fundamental biasing influence when doing research (Clarke, 2005). Or, as Bauman (1992) formulated: "The heuristics of pragmatically useful 'next moves' displaces, therefore, the search for algorithmic, certain knowledge of deterministic chains. The succession of states assumed by the relevant areas of the habitat no agency can interpret without including its own actions in the explanation; agencies cannot meaningfully scan the situation 'objectively', that is in such ways as to allow them to eliminate, or bracket away, their own activity (p. 193)." These questions are a challenge for Human Factors, and Psychology in particular as part of its fundamental scientific background.

1.2   Naturalistic and critical approaches in Psychology

Starting from the 1980s, many scholars left the well-developed, laboratory-based research track to explore alternative paths in attempts to address what they called 'real world' problems. Most prominently, 'naturalistic' approaches were developed to describe cognition 'in the wild' (Hutchins, 1995; Klein, Orasanu, Calderwood & Zsambok, 1993; Zsambok & Klein, 1997). Apart from a fundamentally different orientation towards the generation of empirical material, some of these scholars were - and still are - concerned with the theoretical foundations of Work Psychology and the related Engineering disciplines. They were formulating more fundamental critique towards the underlying world views, especially the concept of technology as such, and they supported the notion of humans as historical, social and bounded-rational actors (e.g. Gigerenzer & Hug, 1992; Parker & Shotter, 1990).

The disciplinary characteristics of Psychology are of great significance as a general background for the scientific project of my thesis. As a scientific discipline, Psychology is an inherently *modern* endeavor that developed much in parallel with industrial automation and the individualistic consumer society of the Western World (Gergen, 1992; Parker, 2007). At the core of modernism lies the belief in a knowable world. Modernist science is striving to discover universal properties, principles and laws that allow for prediction of the events. To achieve that, empirical methods are employed that are believed to be impersonal and free of values or interests. In doing so, modernists believe, scientific progress subsequently leads to the establishment of reliable, value-free truths about the various segments of the objective world.

Starting in the 1950s and 1960s in philosophy, and reaching out into many other fields such as linguistics, literature studies, sociology, anthropology and history, many scholars have provided fundamental challenges to that modernist framework. Taken together, these efforts and their significant consequences for many scientific disciplines have been termed the 'postmodern turn', meaning a shift occurring more or less in parallel in the humanities and social sciences. This turn or shift in thinking encompasses a set of fundamental changes to the philosophical assumptions of these sciences. On one hand, the ontological certainty of a real and potentially knowable world became untenable. On the other, epistemological certainty and the belief in steady progress towards the truth proofed to be the result of the negation of the pervasive influence of social contexts in knowledge production. In departing from a modernist stance, "postmodern thought invites the investigator to take account of the historical circumstances of his/her inquiry. What are the roots of the preferred discourse, what are its limits, what patterns of culture does it sustain, what does it discourage (Gergen, 1992, p. 24)?" As a consequence, *self-critical reflection* is essential for a postmodern scientific approach. This requires "a form of professional investment in which the scholar attempts to de-objectify the existing realities, to demonstrate their social and historical embeddedness and to explore their implications for social life (Gergen, 1992, p. 27)". Therefore, the broad concept of postmodernism as I understand it refers to a kind of scientific orientation, position or style hat distinguishes itself from purely modernist science along the lines sketched out above. In its essence, this would fit to the project that I am writing about in the following. However, I chose not to use the term postmodern because of its many ambiguities and heterogenous use in literature. Instead, I will use the term *critical* to designate emancipatory, historically as well as politically sensitive approaches to scientific work[1].

---

[1] In doing so, I am following Tolman and Maiers (1991) as well as Fox, Prilleltensky and Austin (2009), for example.

Introduction

Questions around the development of an emancipatory, *critical* approach to Psychology have been addressed in theoretical discussions among theoretical psychologists for some time (cf. Kvale, 1992; Tolman & Maiers, 1991; Valsiner & van der Veer, 2000). Fundamentally, there is a substantial amount of uncertainty concerning the methods that are needed to overcome a reductionist psychology that focuses on inter-individual differences rather than the dynamic processes that constitute the individual psyche and its social environment (Valsiner, 2009). But, as Gergen has critically remarked some time ago, there might be a case for the psychological study of *local practices* in defined historical and cultural settings: "(...) while research attempting to accumulate basic knowledge about 'perception', 'cognition', 'emotion', and the like, is of limited value, there remains an important place for sound prediction and personal skills within various practical settings (Gergen, 1992, p. 26)." However, it is still unclear *how* Psychology might contribute to a critical or constructivist social study of technology and its use (Schraube, 2009). Only few psychologists are willing to make the necessary investments and leave the modernist framework to "conjoin the personal, the professional and the political (Gergen, 1992, p. 27)".

Whereas it remains unclear whether Psychology as a discipline will pick up some of the challenges posed by critical thought in sociology and the humanities, Human Factors' more open, interdisciplinary approach might profit from the potential of some of these new ideas by assimilating emancipatory critical thinking and hereby reconsidering at least some parts of its inventory of theories and methods.

1.3 Human Factors in relation to other disciplines

The scope of Human Factors research spans from knowledge-oriented basic research to design-oriented applied research (Vicente, 2000). In basic research, the exclusive purpose that is motivating a study is to contribute to domain-independent theoretical knowledge. In general, this is achieved through vigorously controlled laboratory studies. But basic research has rarely meant to work on the ontological and epistemological fundamentals of the discipline itself, on its basic assumptions and paradigms. It is through design-oriented research efforts that these discussions have been initiated in the past decades.

I understand Human Factors as an interdisciplinary science that seeks explanatory knowledge based on a systemic perspective on the human-technology relationship (cf. Badke-Schaub, Hofinger & Lauche, 2008, p. 7). As such, Human Factors has a *universal/scientific aspect* that is aiming at the generation of knowledge that is domain-independent. Engineering Psychology as a sub-discipline of Human Factors is mainly concerned with basic *psychological* questions arising from the interaction of humans with technology with the aim of generating universally applicable scientific knowledge. Human Factors furthermore has a *design-oriented aspect* that is known as Cognitive Systems Engineering. Those two aspects could also be characterized as micro- and macro-psychological orientations and methodologies within Human Factors, where Cognitive Systems Engineering is representing the macro-psychological branch (Zimolong, 2006, p. 17f). Table 1 is providing an overview of these two aspects.

Human Factors as a discipline, hereby including Engineering Psychology and Cognitive Systems Engineering, employs a specific *concept of technology* which could be described as essentialist and inherently modernist in nature. To address the questions raised by large-scale information technology systems that are widely distributed and dynamic in nature, it becomes a necessity to extend that concept. The question remains open if one of the two main aspects of Human Factors described above will transform itself to accommodate these new technologies or we need a third, more or less independent approach with its own theoretical base and preferred methods.

The motivation and broad intent of my thesis is with making a contribution to the understanding and design of such new, highly networked and distributed systems. To achieve that I have to recur to fundamental assumptions and principles of our science and examine them, if necessary with reference to Philosophy (cf. Clegg, 2000, p. 464), as I will show in detail. My intention is further to use literature from neighboring disciplines such as Psychology, Sociology, Management Studies, Organization and Information Sciences to put my propositions into a context that reaches beyond Human Factors.

Table 1: Characteristics of two important aspects of Human Factors as a discipline

Aspect	Universal/scientific	Design-oriented
Label	Engineering Psychology	Cognitive Systems Engineering
Orientation	Micro-psychological	Macro-psychological
Root discipline	Experimental Psychology	Industrial and Mechanical Engineering
Objective	Domain-independent knowledge generation	Accepted designs for specific domains or purposes
Paradigm	Information processing	Systems theory
Approach	Systematic-experimental, controlled, observation	Analytical-pragmatic, case study, observation, qualitative methods
Favored methods	Mathematical Modeling, Experimental Testing, Statistics, Simulation (micro-worlds)	Abstraction Hierarchy, Ecological Interface Design, Critical Decision Method, Simulation (micro-worlds), Mathematical and Computational Modeling
Scholars	Klix, Timpe, Hacker, Dörner, Wickens, Manzey, Klein, Parasuraman, Fogg	Rasmussen, Moray, Vicente, Hollnagel, Woods, Burns, Jamieson, Naikar, Sanderson, Kirlik

In doing so, I follow others who have attempted to combine thoughts from different perspectives to work on innovative approaches to technology and organizations (cf. Erlicher & Massone, 2005; Leonardi, 2012). Most of them, including me, are trying to establish pragmatic – but theoretically well rooted - relationships between principles, design methods and criteria to promote and undertake sociotechnical design (Clegg, 2000; Mutch, 2013; Orlikowski & Scott, 2008; Rammert, 2007). My wish is to bring Human Factors, organizations and society in general closer together. I am convinced that, in considering critical thought within our work, technology will be placed more directly and openly in the service of values. In the process, we as scientists are "encouraged to join in forms of valuational advocacy, and to develop new intelligibilities that present new options to the culture (Gergen, 1992, p. 28)".

1.4 Contents of thesis

Following these considerations, my dissertation is concerned with the discussion of constructivist, emancipatory or critical approaches to technology studies and, accordingly, the possibility of a 'critical' Human Factors approach. Drawing on field research in computer-supported industrial production planning and scheduling, I explore and discuss the validity and potential of a selection of self-critical, constructivist achievements for Human Factors as an application-oriented interdisciplinary science. The aim is to critically discuss and evaluate paradigms of qualitative research approaches to human-technology problems. The thesis is primarily intended to contribute to a scholarly discussion about the fundamental *philosophical assumptions* of Human Factors and technology studies, and secondarily to propose theory-guided *analytical approaches* and methods as possible extensions of today's Human Factors methodology.

In the following chapters, I therefore lay out the problematic relationship of Human Factors theories, models and design principles within production planning and scheduling as an *example* of a complex computer-mediated work context. Hence, in chapter 2, the field of industrial production planning and scheduling is introduced, including a description of the tasks that are involved and the technology that is employed to support them. Chapter 3 is then summing up the Human Factors research that has been conducted in this field. It includes a discussion of the limitations and challenges of such approaches to address some of the most pressing Human Factors issues related to the implementation and use of production planning technology.

Chapter 4 introduces a general overview of the history and actual situation of Human Factors as a scientific discipline and discusses possible critical, constructivist or emancipatory theoretical perspectives that could be adopted by the discipline to cope with the highly complex research domain described in chapter 3. Hereby I advocate a post-essentialist view of technology and an activity-oriented interactionist perspective in order to overcome the weaknesses of the current Human Factors approach to complex socio-technical work environments. I am reviewing some existing work that – in my opinion – is opening up pathways in this direction. A main focus lies on the discussion of ontological and epistemological fundaments of various theoretical approaches that might be employed. After a philosophical re-orientation towards *critical realism* as a framework for a social science that acknowledges a *a priori* reality, I develop an analytical approach that is in accordance with this framework. I then discuss potentials and implications of that theory-based approach to the analysis of technologies, social structures, intentionality and related discourse in complex, intentional work domains.

To support the theoretical discussion in chapter 4, I analyze and discuss empirical material from field work in two manufacturing companies in chapters 5 and 6. Chapter 5 consists of a critical analysis of interview data as texts that are reflecting socio-cognitive discourses. This shows how technology discourse and work domain characteristics are interrelated. Post-implementation adaptation and actor strategies are made explicit using these methods. The study is innovatively linking Cognitive Work Analysis to an early-phase requirements engineering method, the *i** modeling framework. Chapter 6 builds upon the insights gained in the first study to analyze field data from a second study using an extended and adapted version of Cognitive Work Analysis as well as Archer's morphogenetic approach (Archer, 1995; Mutch, 2010) to further illustrate and test the proposed methodology.

Chapter 7 is dedicated to the discussion of the case study results as well as some conclusions relating to the conceptualization of agency and social structure in Human Factors. I then critically discuss how the results from applying the proposed theory/methods package to concrete cases can contribute to improve and further elaborate this package.

In the final chapter, three kinds of implications of the findings provided by the two studies on production planning in manufacturing organizations are sketched out. The first set of implications is on the theoretical level, concerning the possibility of a critical realist Human Factors approach and the issues involved in the role of practitioners when working with organizations. The second set of implications is concerning practical questions of research and development, and the final set is concerning teaching and training. I finish my thesis in chapter 8 with an outlook on possible future directions in Human Factors research related to large-scale information systems in organizations.

2 Computer-supported planning and scheduling in manufacturing

> Particulars are important for theory building, and theory is important for making sense of specifics (Orlikowski & Barley, 2001, p. 147).

When we talk about planning, we generally distinguish between planning as 'planning for oneself' versus planning 'for an organization' (van Wezel & Jorna, 2001). Planning for oneself has been studied widely in cognitive psychology, artificial intelligence and robotics. It spans action regulation as well as problem solving (Hoc, 1988; Morris & Ward, 2005). As opposed to that, planning within organizations is a complex task that is aiming at the coordination of distributed activities that involve others. It is process-oriented, requires the representation of a plan and is often reversible within a certain time frame (Jorna, 2006, p. 111). Only the second is of interest here, although the first is naturally involved in the work of the individual industrial planner as a working professional. Production planners are doing both, planning their own behavior as well as planning and scheduling the activities within their field of responsibility (Resch, 1988). Planning takes place within 'centers of coordination' (Suchman, 1997) that can be characterized as follows: "Centers of coordination are characterizable in terms of participants' ongoing orientation to problems of space and time, involving the deployment of people and equipment across distances, according to a canonical timetable or the emergent requirements of rapid response to a time-critical situation (Suchman, 1997, p. 42)".

Planning and scheduling as an activity that strives to achieve coordination addresses diverse organizational dependencies such as shared resources, producer - consumer relationships, simultaneity constraints and task - subtask hierarchies (Crowston, 1991; Malone, Crowston & Herman, 2003). In the next section I will present some overall purposes and organizational constraints that are shaping production planning in most manufacturing companies. Then I will describe the tasks of planners and schedulers on different levels of a prototypical organization. And in the last section I will briefly describe the tools that are currently used to achieve these tasks. These chapters serve the introduction into the field of production planning and scheduling, in order to prepare for the critical discussion of current human factors research in the subsequent chapters.

2.1 Industrial production planning and scheduling

Historically, production planning and scheduling became more important in the age of industrialization. Production engineers like Frederick Taylor and Henry Gantt developed important methods and techniques that are still relevant today (Herrmann, 2007). These methods were designed to cope with the increasing complexity of production due to the division of labor along the supply chain. Orders have to be fulfilled in time, they usually come with a due date. Orders are competing for resources, especially highly skilled workers or expensive high-tech machinery with limited capacity. Stock levels have to be kept low in order to avoid capital lockup. But nevertheless delivery reliability has to be kept high in order to satisfy customers. In addition to that, a wide range of operational uncertainties have to be dealt with day by day. Raw materials might be scarce or their delivery delayed. Construction plans and other documentation may be faulty or unavailable. Tools may not be ready, key persons such as decision makers or skilled workers may be absent, or communication between departments may be difficult.

The introduction of information technology into this complex work domain arrived in parallel with increased efforts to create more flexible manufacturing systems. Hereby, management principles, organizational structures and operational control efforts are closely interlinked and relying not only on formal but also informal structures. Centralized control and standardization is in conflict with the need for local flexibility on the workplace level. Intimate knowledge of the work process is necessary to react to operational uncertainties and create local solutions. Perceived from a distance, and with powerful computers available to crunch enormous amounts of data, the planning problem is frequently seen as a problem of not enough measurement and data rather than a question of dynamic and adaptive control strategies that are also requiring some decision latitude (cf. Rasmussen et al., 1994, pp. 2-6).

A simple input-output model of production is shown in Figure 1. Around the 'black box' of the production process in the middle there are various dependencies between the work system and its environment. The production or - in more general terms - transformation process depends on timely delivery and adequate quality of raw materials and purchased parts, it depends on energy delivery and information relevant to the transformation that is intended. The transformation process[2] further depends on existing facilities, available personnel and know-how. It is disturbed by the environment, be it unfavorable weather conditions or other unexpected events. The process itself produces unexpected behaviors that are to be coped with by the environment. It further produces the product and/or services that are desired by the customers, plus waste that is either sold to other companies or left to the environment to be dealt with. Planners and schedulers are concerned with all these aspects of the transformation process. This somewhat simplified view of manufacturing has been criticized lately, since it neglects important dynamic and social aspects of manufacturing (cf. Ahrens, 1998). Planning and scheduling in itself is a process within the work system but is not identical to it, nor can it be easily located and demarcated. As opposed to the transformation process as the 'primary process' it has been called a 'secondary process' as it fulfills a support or control function of the primary process (Wäfler, 2001).

The understanding and optimization of planning as secondary or control process of manufacturing has been a research target for many decades and in different scientific disciplines. Mathematicians within the operations research domain were investigating optimal solution algorithms for complex problems with solutions spaces too large for computation that are raised in production scheduling (cf. Domschke & Drexl, 2007; Parunak, 1991; Simon & Newell, 1958). Business engineers were developing organizational structures and tools to cope with more and more demanding and increasingly interrelated

---

[2] Dessouky, Moray & Kijowski (1995) use the term process - *transfer function* to lable the production process at the core of the illustration. I prefer to use the term *transformation process* (in German Transformationsprozess or Leistungserstellungsprozess, cf. Schuh, 2006). Furthermore, instead of using *equivocation* I use *unexpected behavior* to describe system outcomes other than products, services and waste. Dessouky and his colleagues are not labeling the system itself, I chose to call it *work system* since its purpose is the skilled, effortful and intentional creation of a utility or commodity.

operational management tasks (cf. Schneider, Buzacott & Rücker, 2005; Schönsleben, 2004). Work psychologists and cognitive engineers were interested in the mental models of planners and schedulers to understand human planning behavior (cf. Sanderson, 1989; Strohm, 1996) and to provide computer engineers with a 'model human scheduler' (Sanderson, 1991). They also investigated the tasks, roles and cooperation strategies planners are using to accomplish their objectives (cf. Jackson, Wilson & MacCarthy, 2004; Kellogg et al., 2006). Industrial sociologists were analyzing the social relations within production systems and the shaping of agendas and practices, as well as technologies, through these power structures (Hill, 1981; Noble, 1984; Webster, 1991b).

Whereas most of these scholars agree on the basics of the transformation process as shown schematically in Figure 1 - however limited in perspective it is, they have diverging opinions on the nature and dynamics of the secondary process that controls it. A variety of control theoretic models of production planning exist - depending on the theoretical background or the intentions of the authors. The most basic formulation of such a model is based on the classical control model of engineering, where an output signal is monitored and a feedback mechanism ensures a modification of the input signal (e.g. thermostat).

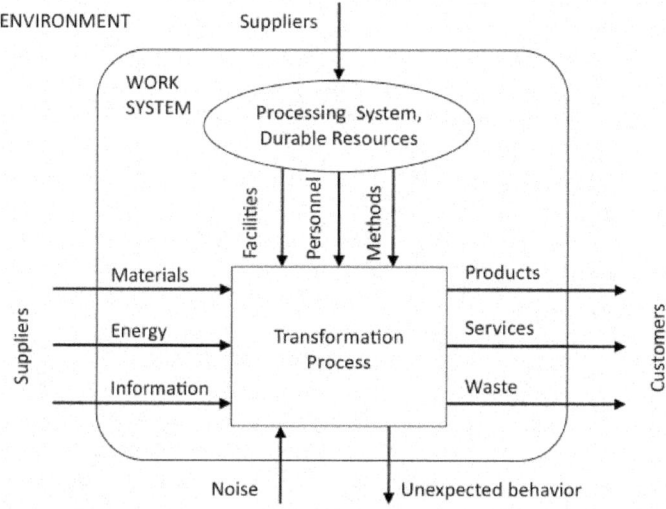

Figure 1: Characteristics of generic industrial transformation process (adapted from Dessouky et al., 1995, p. 453)

One basic distinction that extends the 'engineering model' is involving the notion that any kind of control in a dynamic environment must include some form of learning. As Argyris observed many years ago (1977) in his seminal article on *organizational learning and management information systems*, a learning organization is based on a double loop system of goal setting and measurement. Figure 2 is illustrating these processes that are constitutive for production planning and scheduling. It is based on a framework proposed by Kontogiannis (2010b).

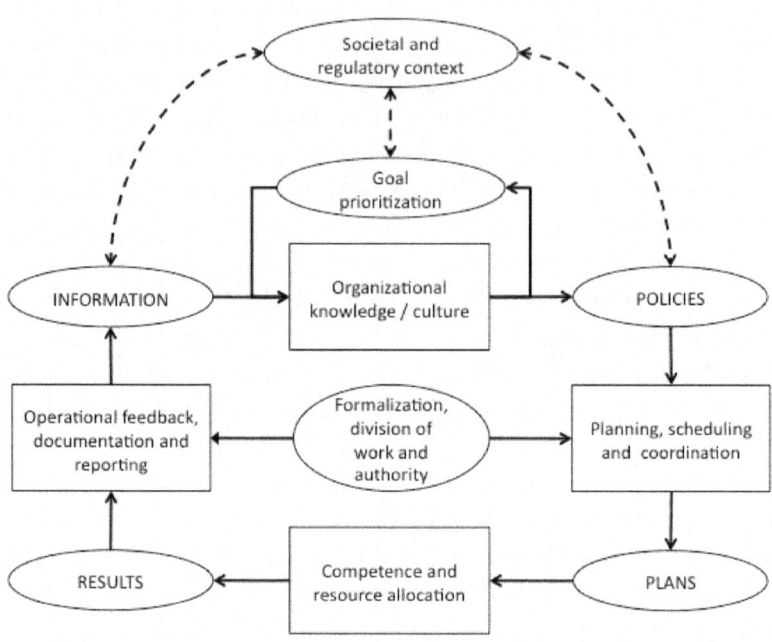

Figure 2: Generic framework of production control (adapted from Kontogiannis, 2010b)

The upper part of Figure 2 is showing the dynamics related to goal prioritization and organizational knowledge. The wider societal context is highly influential when it comes to issues like workplace safety, shift work or other work-related regulations. Depending on the organizational culture, production priorities may be set in terms of quality, cost or timeliness. These factors influence the formation and implementation of organizational policies. The lower part of the schema describes the dynamics surrounding the actual production process. There is a basic feedback loop involved in planning, resource allocation, operations, feedback and adaptation of the policies and planning practices. The crucial point in my perspective is the division of authority which is depicted in the middle. In this illustration, it seems set and invariable. There are no dynamic influences that would change the general distribution of work and control within the system. Since the introduction of powerful and highly accessible information systems in organizations, it seems unlikely that flexible and dynamic companies would not try to optimize this central element of control. In fact, many are constantly re-engineering their control tools and processes.

This is why my colleagues and I have proposed a different perspective on organizational control capabilities that involves a differentiation between structural control and operational control (Wäfler et al., 2008; Wäfler et al., 2011). Both control modes are based on a specific distribution of authority within an organization. Structural control usually is allocated within the higher management, since it involves the capability to change control structures and encourage autonomy on all levels of the organization. In highly decentralized organizations however, structural control may be located at many levels, including the shop floor. Figure 3 is showing a formal representation of this control model.

In general, control capability determines the possible control behavior of an organization's employees. Control behavior may or may not exploit the full potential of control. Not exploiting control capability means that the employees – for one reason or the other – do not utilize the scope of behavior that would be possible according to their skill level. Consequently, as our model in Figure 3 shows, control behavior is a function of control opportunities as well as control skills on the one hand and control motivation on the other hand. There might be more control opportunities than can be used by the workers in a work system, due to a lack of knowledge or training. In our model we further distinguish two modes of control behavior: *Operational* control behavior refers to all attempts to achieve control without changing the structure of an organization. *Structural* control behavior refers to control by changing on organization's organizational or technological structure, e.g. changing the shift model, changing production processes, implementing new monitoring tools or qualifying employees.

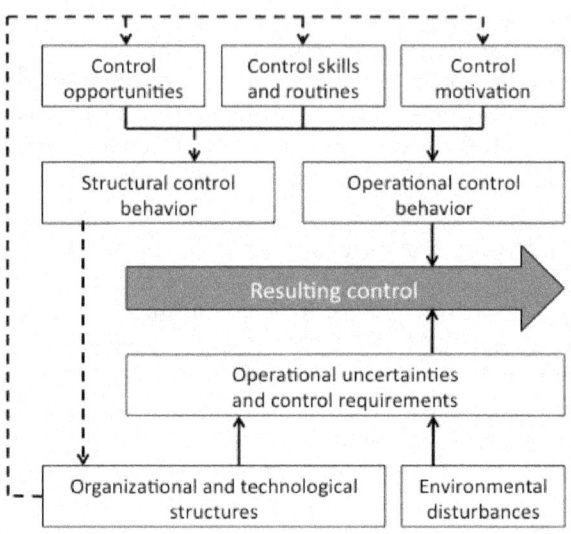

Figure 3: Structural and operational control behavior (adapted from Wäfler et al., 2008)

The organizational and technological structure facilitates or restricts control opportunities, control skills and control motivation. This feedback loop is depicted using dashed lines, to distinguish it from the operational control loop. The organizational and technological structure plus environmental disturbances are the main sources of operational uncertainties and control requirements. This model makes clear that control behavior in organizations is dynamic in nature, and not only determined by slowly changing policies and mostly static organizational structures (cf. Figure 2). In many industries, flexibility is increasing due to more decentralized information systems, and organizational structures become more and more temporal through ad-hoc team building, project structures or semi-autonomous groups of workers. Therefore, a perspective of engagement and participation in control issues becomes more important and potentially significant for the overall performance of a work system. Are workers interested and motivated in shaping and improving control structures and practices or are they merely following prescribed procedures?

In most manufacturing companies production planning is achieved through the use of so called enterprise resource planning (ERP) systems that have been introduced in a general industrial change process towards a more computerized and standardized mode of production (Dubois, Heidenreich, La Rosa & Schmidt, 1995; Umble, Haft & Umble, 2003). These technologies are highly complex socio-technical systems that are - at the core - based on material requirements planning (MRP) algorithms. However, because of severe limitations of the accuracy of underlying software models, internal and environmental dynamics and general data quality issues, it is indispensable - for most companies - to employ highly experienced planners and schedulers that are able to cope with these weaknesses. These human planners are facing a wide range of difficulties due to system complexity, intransparent automation, lack of anticipation, informal social dynamics as well as rigid formalized procedures. Most EPR systems were designed with an essentialist conception of the technology ('technology-driven' product development) and are based on algorithms that were developed by mathematicians - regardless of company-specific variations in purpose and use of this technology. This gap has led to many failures in ERP implementations at a substantial cost for these organizations (cf. Shepherd, Clegg & Stride, 2009; Umble et al., 2003).

One could argue that these failures are not only due to a general gap between standardized software products and day-to-day reality in manufacturing, but also due to various human factors issues, ranging from data management to managerial practices and standards. In order to approach the planning and scheduling work domain from a Human Factors perspective, it is therefore necessary to understand the most important tasks in planning and scheduling, regardless if accomplished through the use of computers or not. Furthermore, it is necessary to understand the most common division of labor in production planning and the basic functionality of computerized tools and their underlying standard models.

2.2 The planning and scheduling task

The purpose of planning and scheduling in manufacturing is to mediate between market demand and available production resources. The task - and at this level it might be better to speak of an organizational *role* - requires a wide array of skills and competences (Berglund, Guinery & Karltun, 2011; Jackson et al., 2004). Planning tasks are nested within organizational hierarchies, are recurrent by nature and demand substantial cognitive efforts usually involving industry-specific expertise.

Significant amounts of research in planning and scheduling have been undertaken since the 1950s (cf. Crawford, 2001; Sanderson, 1989). Analytical models, heuristics and scheduling hierarchies have been developed and investigated, mainly in Operations Research and Artificial Intelligence communities. All these concepts had relatively little impact on the everyday work of schedulers due to a myriad of uncertainties in 'real world' operations that no model or algorithm is able to account for. It remains the task of the human scheduler to integrate expert knowledge about the work system into the generation of a feasible schedule (McKay, Buzacott & Safayeni, 1989). Basically, his or her task is "to schedule and dispatch work in such a way that many stated and unstated conflicting goals are satisfied using hard and soft information that is possibly incomplete, ambiguous, biased, outdated, and erroneous (McKay et al., 1989, p. 173)".

Given this inherent vagueness in the planning and scheduling task, there are nevertheless general features that can be described. Planning activities take place in a *hierarchical structure* that constitutes itself through the necessity of long-, mid-, and short-term planning. They furthermore consist of *repetitive* tasks due to the *temporal cycles* of industrial production. Within this basic setting, planning and scheduling tasks can be described on a local level in *specific contexts*. Such descriptions necessarily have a limited scope that makes generalizations difficult. Therefore Hoc (1993) and Cegarra (2008) have suggested a more formal approach to describe situational characteristics by using cognitive typologies for process control and scheduling situations in general. In the following sub-chapters I summarize these distinctive features of the planning and scheduling task, in order to introduce the work of the industrial planner. In a subsequent chapter I then present the tools that are currently in use for planning in manufacturing.

2.2.1 Planning hierarchy

From an organizational perspective, the planning task is typically distributed over a hierarchical structure reaching from long-term system planning to mid-term order release to short-term personnel dispatching and machine scheduling. A prototypical structure would involve five interconnected levels of planning and scheduling (cf. McKay, 1992a; McKay & Wiers, 2006; McPherson & White, 2006):

1. System planning: Determines the manufacturing resources needed to achieve long-term goals, e.g. decisions about equipment, employment policies and reporting requirements.
2. Production planning: Establishes production rates for aggregate product classes, considering market demands and production capacities.
3. Flow planning: Determines batch sizes, process steps and flow times for each product.
4. Scheduling: Decisions about the implementation of the various flow plans in a coordinated and consistent way through sequencing and timing.
5. Cell control and machine control: Shop floor level dispatching of jobs, material and operators as well as operation planning in general, e.g. set-up of machine, control, maintenance.

The various tasks in planning and scheduling are mostly spanning across several of these hierarchical levels. As McPherson and White (2006) have described, there are two important aspects to be acknowledged when talking about these interactions: First, the distinction between planning and operations is often blurred when we are describing the achievement of a plan. In order to be successful, the planning activities on superior levels must be consistent with control capabilities at subordinate levels, and planning at subordinate levels must be consistent with the overall goals of the work system. In other words, control opportunities must exist on subordinate levels, and they need to be exploited in a way that contributes to the overall objectives (Wäfler et al., 2011). Second, the degree of control a subordinate decision maker can achieve over sources of uncertainty depends on local control opportunities and skills. Goal achievement may or may not be feasible within a specified time horizon.

These two fundamental characteristics of hierarchical interactions within planning tasks are constitutive for dynamic reactions within manufacturing organizations, which could be formulated as bottom-up and top-down effects that lead to a specific kind of *resonance* within the decision hierarchy. From the bottom up, subordinate levels may defer planning decisions due to infeasibilities on local grounds. This causes reactions at superior levels, depending on the amount of decision autonomy that is allocated to subordinate levels. Top-down adjustments of commands then in turn cause reactions at subordinate levels. This can cause an immediate bottom-up reaction, which could be called *resonance*. This reaction enforces consistency among the goals of the other levels. Both dynamics are rather slow and information-intensive (McPherson & White, 2006; McPherson & White Jr, 1994). Moscoso, Fransoo, Fischer and Wäfler (2011) have described a similar dynamic phenomenon which they call the *planning bullwhip effect*, referring hereby to the bullwhip effects in supply chains.

Alternatively, non-hierarchical models of organizations and planning have been proposed (cf. Ahrens, 1998; Mehnert, 2004). These models have the advantage that they contribute to an understanding of industrial organizations as communities or networks that have multiple layers and are characterized through social, political, technological as well as economical dynamics. In this view, planning does not only take place in a seemingly hierarchical force field of managerial control versus worker autonomy. Organizations are perceived as configurations of self-referential entities that are nested and capable of self-organization. Others have taken this approach further and are discussing networks of social and technological actors or agents that are acting towards goal achievement (Latour, 2005; van Eijnatten & van der Zwaan, 1998).

A hierarchical view of the organization implicitly evokes a linear model of time in planning. This is greatly misleading, since production planning and scheduling never actually 'starts' or 'ends', it is always ongoing, cyclical, and a never-ending variation of routine tasks.

2.2.2 Temporal cycles and recurrent tasks

A significant characteristic of planning and scheduling tasks is the presence of temporal cycles that are more or less synchronous to each other, meaning that they are regular to a certain degree, or out of pace altogether (cf. Cegarra, 2008). Planning and scheduling is an ongoing process that involves the same routine tasks at specific time points or within irregular intervals. It is highly responsive to disturbances, which are often leading to re-planning (opportunistic or adaptive planning; cf. Alterman, 1988; Hayes-Roth & Hayes-Roth, 1979).

To facilitate the description of scheduling tasks in any given environment, Dessouky, Moray and Kijowski (1995) have proposed a taxonomy of scheduling that is based on a classification of scheduling *problems*. They first distinguish between single and multiple stage scheduling problems. If there is only one stage, the job consists of several tasks. When the problem consists of multiple stages, the authors distinguish between the job (mission) and its component operations, where each is again composed of different jobs and tasks. Parameters are entities like processing times, due dates, setup times, priorities, precedence, preemption, repetitiveness, efficiency, machine capacity, availability, delays and other system characteristics. In addition to these problem characteristics the authors provide a list of objectives and criteria that allow for the overall performance of a plan or schedule to be evaluated over time. These objectives include productivity, delivery reliability, workload management and other specific objectives like throughput or resource utilization (Dessouky et al., 1995, p. 470).

An isolated view of scheduling problems as cognitive tasks or problems that need to be solved ignores the cyclic nature and sensitivity to disturbances of planning and scheduling activities. Therefore, a formal taxonomy only describes certain aspects of the work that is done by planners and schedulers. To overcome this limitation, several field studies have been conducted to investigate the cognitive work that is performed within planning departments.

2.2.3 Task analysis of planning and scheduling

Planning and scheduling tasks are important many work domains. This includes for example cooking in a restaurant kitchen (Fine, 1990), nursing in a hospital ward (Wolf et al., 2006), or ambulance dispatching (Furniss & Blandford, 2006). Despite the pervasiveness of the task as such, there are not many scientific contributions devoted to the analysis of cognitive work related to scheduling. There is relatively little practice-oriented, ethnographic work to be found in literature addressing planning and scheduling in the 'real world' (e.g Crawford, 2001; Farrington-Darby, Wilson, Norris & Clarke, 2006; Karltun & Berglund, 2010; McKay, 1992b; Wäfler, 2002). Motivated through this lack of analysis of the planning practice, and the difficulties in distinguishing between the various tasks involved, McKay and Wiers (2003) are describing planning, scheduling and dispatching activities by characterizing them using three aspects: (1) Horizon and timing, (2) decision making and (3) context. Based on these descriptions, the authors suggest that the tasks can be separated by identifying the nature of the input and output. If the input is based on expectations of the future, then the task is planning (and sometimes scheduling). If the output is formulated in orders, assignments and timed jobs, then the task is scheduling and dispatching. The authors acknowledge that task analysis in planning and scheduling has not been satisfactory in the past, and that in industry one rarely finds the 'school book' version of planning and scheduling within a neat and well-structured hierarchy. According to them, the following weaknesses can often be found: (a) The difference between planning and scheduling is not clear, (b) dispatching is underestimated and (c) the scheduling system is intransparent and the level of software support is too high.

Usher and Kaber (2000) used goal-oriented task analysis to identify information requirements in manufacturing system control. They restricted their analysis to the level of job shop scheduling, therefore omitting superior or subordinate levels such as planning or dispatching. Their analysis did not involve actual field work, it is based on a description of a manufacturing facility. Through their analysis they identified four sub-goals that consist of meeting due dates, avoiding bottlenecks, expediting critical orders and maintaining normal system functioning (cf. Figure 4, further below). In order to achieve these sub-goals, several tasks are required in part or in sum. It is interesting to observe that these task involve not only problem solving, but also monitoring and identification or diagnosis. The authors are referring to these tasks as 'objectives', which makes it somewhat unclear if they are talking about tasks or goals. Criticism of the method or application of it aside, this analysis is interesting for the discussion of planning and scheduling task analysis in general. It shows specific shortcomings of such efforts, mainly the artificial boundary-setting in between work domains. In this case, the scheduler seems to work independently of the planner or the dispatcher. This is rarely the case in manufacturing companies.

McKay and Wiers (2006) are therefore describing the organizational interconnectivity in planning and scheduling. None of these tasks can be considered in isolation. They distinguish between structural and functional interconnections and are discussing six aspects or dimensions of these. An overview is provided in Table 2.

Information visibility as a structural interconnection can be high or low depending on the distance of the planning timeframe. The depth of decisions is depending on how far the consequences of a decision reach in terms of the product portfolio. The width of a decision then is defined as the stretch of the supply chain that is affected by it, be it a single cell or machine or reaching beyond the factory up- or downstream of the supply chain.

Table 2: Overview of structural and functional interconnections in planning and scheduling (adapted from McKay & Wiers, 2006)

Structural interconnections	- Information visibility
	- Depth of decisions
	- Breadth of decisions
Functional interconnections	- Information flow
	- Scope and formalism
	- Solution space

A functional interconnection is originating in the amount and quality of the raw data to be acquired and disseminated for scheduling and planning. A second functional interconnection is the scope and the formal or informal nature of a planning task. Finally, the solution space can be rather limited or very large, with many options and situations to consider. All these interconnections relate planners and schedulers to other functions in the organization, from sales to purchasing, and from engineering to quality management. They are not easily located in a hierarchy, they depend on and exploit formal as well as informal structures within the organization. According to McKay and Wiers, an understanding of these interconnections can possibly contribute to future explanations of successes and failures in planning and scheduling technology.

Akkerman and van Donk (2009) compiled a list of possible planning and scheduling subtasks using three different studies from literature. They found 12 such tasks: Assigning, selecting, ranking and counting jobs, monitoring performance, estimating results, administrating production, interpreting data, communicating schedules, evaluating actions, investigating and reacting to events. Interestingly, again, as in the study of Usher and Kaber (2000), the tasks involve seemingly 'unrelated' activities like interpreting data, evaluating actions and investigations.

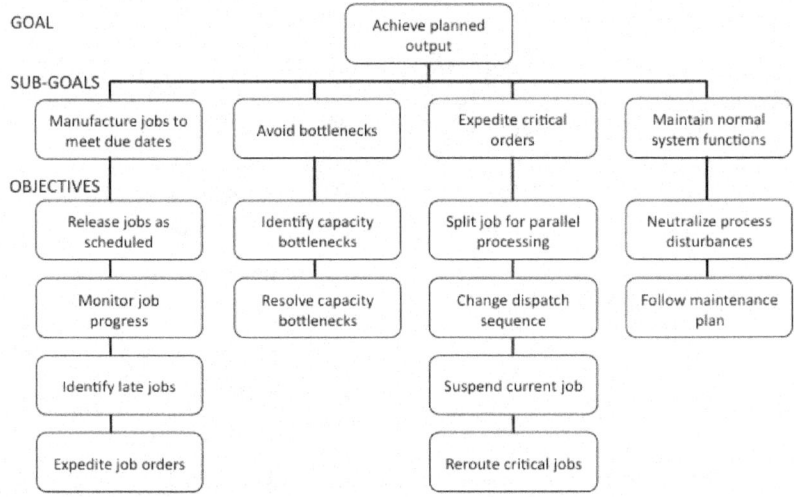

Figure 4: Goal-oriented task analysis of supervisory production control in a flexible manufacturing environment (adapted from Usher & Kaber, 2000, p. 440)

But as others have observed, these are only some of the characteristics of planners in their role within the organization. There is no general agreement on the exact nature of the 'scheduling function' as it is performed by humans. As Crawford and Wiers (2001, p. 44) put it: "(...) there is no definitive statement about what the scheduling function actually entails (...). Indeed, this signifies that there is no one correct way to study scheduling. Researchers and practitioners must ensure that they do not hold too narrow a view of the scheduling function and its position within an organization. Assumptions as to what is meant by the terms 'planning' and 'scheduling' can only constrain the domain of the human factors of planning and scheduling. The goal for researchers must be to challenge these previous assumptions and produce valid and practical definitions and applications that can be utilized by researchers and practitioners alike."

As I have been able to observe and document in our own field work, planners and schedulers are moreover devoting a substantial amount of time in making sense of ambiguous data, building relational knowledge and maintaining data quality and consistency (Gasser, Fischer & Wäfler, 2007; 2011). These tasks are not directly related to planning and scheduling in the narrow sense, but day-to-day planning and scheduling would not be feasible without them. It thus is a knowledge-intensive activity which does not only involve know-how about scheduling as a discipline, but also a great deal of knowing about organizational and technical processes; and to what extent those can be administered using highly elaborated computer systems. Most planners we have observed and interviewed were well aware of strengths, shortcomings and limitations of their digital tools (see also Bechky, 2003; Karltun & Berglund, 2010). According to Cegarra and van Wezel (2010a) this is due to the predominant use of normative task analysis methods (like in the example above; Usher & Kaber, 2000) which often leads to situations where planners have to adapt to their computer systems - or are using external tools to complement and support the work with the ERP system. Cegarra and van Wezel are therefore suggesting to use a combination of task analysis methods according to the focus and scope of the investigation.

2.3 Enterprise resource planning systems

Historically, there is a wide range of planning tools and techniques (Herrmann, 2005; McKay & Wiers, 2003). In the last decade, driven by the accelerated globalization in production, there has been a trend towards the use of relatively standardized manufacturing control systems. These so-called enterprise resource planning (ERP) systems are gigantic infrastructures that link functionally disperse, geographically distributed and culturally different organizational units into a uniform system (Shehab, Sharp, Supramaniam & Spedding, 2004; Vollmann, Berry, Whybark & Jacobs, 2005). ERP has been defined as: "... a method for the effective planning and controlling of all the resources needed to take, make, ship and account for customer orders in a manufacturing, distribution or service company (Møller, 2005, p. 485)".

The hope and promise of ERP systems is that they provide common data, common procedures and real-time data availability for coordinated decision making. Originally, manufacturing control systems were designed to support individual factory operations. ERP allows firms to move beyond this boundary and to integrate several factories as well as functions such as accounting, human resources and sales. More and more they are also aiming at optimization of the supply chain, and are therefore reaching across the boundaries of the company.

Most of todays ERP systems are based on the MRP-II framework (MRP stands for 'materials requirements planning', and the number II is indicating the use of capacity constraints in addition to material requirements, cf. Møller, 2005). The main features of this framework are depicted in Figure 5. The core functionality is material requirements planning, which involves part lists ('bills of material') and operation plans for each product to determine timed material requirements as well as stock levels and replenishment times to calculate re-order points for specific assembly parts or raw materials. Depending on the mode of production, internal and external production and purchasing orders are released. Typically, material requirements planning is updated on a daily or weekly interval, but there are existing systems that allow for real-time re-planning, and even the simulation of scenarios. These are called 'advanced planning systems' (APS). While the global trend goes toward uniformity, local differences in the design and implementation of ERP systems have been reported (Porter et al., 1996). Interestingly, there still are many local and comparably small providers of ERP systems. As Møller (2005) reports for 2001 and 2002, there was a 50% market share divided among the six biggest providers of ERP systems, leaving the other half to these many small or specialized software companies.

There has been a substantial amount of research on information systems designed for production planning in the past decades. The most recent review by Schlichter and Kraemmergaard (2010) reports on 885 peer-reviewed publications on ERP systems for the period between 2000 and 2009 and a review by Moon (2007) included over

300 articles from within just six years up to June 2006. Another review found 189 papers from 2000 to mid 2004 (Møller et al., 2004). The numbers indicate an increasing research interest in ERP systems during the last decade (Schlichter & Kraemmergaard, 2010).

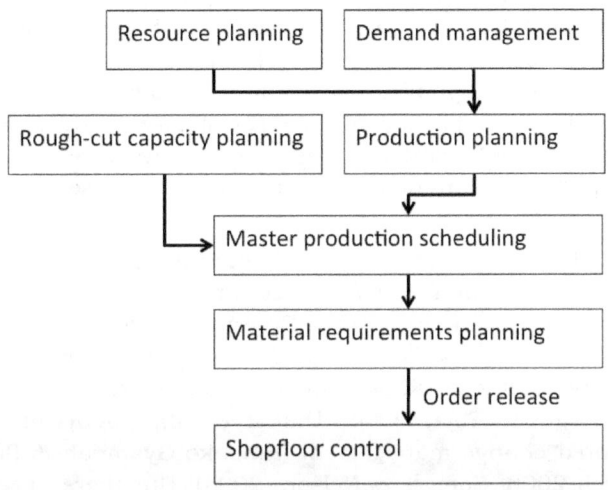

Figure 5: MRP-II framework (adapted from van Wezel & Jorna, 2001)

Shehab, Sharp, Supramaniam and Spedding (2004) have reviewed 76 papers on ERP technology from 1990 to 2003. Their main conclusions are that ERP systems are costly and difficult to implement, especially for small and medium enterprises. However, the pressure is rising due to the trend that big companies are integrating their whole supply chain into one ERP platform. The authors are providing a list with the main flaws of ERP systems that they have found in the literature: Many functional, technical and usability-related issues are troubling companies that are implementing these systems.

In their review of 49 articles that were focussing on ERP system research, Cumbie, Jourdan, Peachey, Dugo and Craighead found a high prevalence of exploratory and case studies, whereas confirmatory research designs were less often employed. They conclude that "to advance the field of ERP, researchers need to continue to explore creative, multi-method research to overcome the inherent complexities when studying an enterprise (2005, p. 29)".

Schlichter and Kraemmergaard (2010) conducted the most thorough and comprehensive review up to date. Their study shows that a large number of journals have published papers about ERP. The Operations Management discipline has hereby published 31% of the papers, followed by Information Sciences with 24%. The most researched topic has been the study of implementation of ERP, accounting for 29% of the publications. Studies on the management of ERP (18%) and the optimization of ERP (17%) together account for another third of the publications. Case studies have been the most used method (22%) but in more recent years their use has been declining.

Within much of the existing research literature, the main focus has been on implementation issues and best practices (Akkermans & van Helden, 2002; Davenport, 1998; El Amrani, Rowe & Geffroy-Maronnat, 2006; Finney & Corbett, 2007; Holland & Light, 1999; Hong & Kim, 2002; Ifinedo, Rapp, Ifinedo & Sundberg, 2010; Klein, Conn & Sorra, 2001; Soja, 2006; Umble et al., 2003) and ERP technology related change management (Amoako-Gyampah & Salam, 2004; Shepherd, 2006; Somers & Nelson, 2004). But there have also been studies around other aspects of ERP application, for example on user interaction, team collaboration and local optimization (Calisir, 2004; Cegarra, 2008; Gattiker & Goodhue, 2004; Tan et al., 2008; Wallace, Scott, Stutz, Enns & Inkpen, 2009; Windischer, 2003).

ERP systems depend on a quite detailed model of the production or transformation process and all the resources that are involved. Some of these systems are linked to the wider supply chain, leading to more and more integration of data flows. Parts of the system are automated, which in some companies lead to the expectation that all of the production planning can be automated and would only need little human supervisory control some time in the future. Such hopes of 'effortless' supervisory control in advanced manufacturing systems are misleading. In many companies, day-to-day planning and scheduling of operations still is a challenge despite the increasing powers of manufacturing control systems. Critics have long made the point that - when not very carefully managed - ERP implementations can lead to the situation where the organization has to adapt to the software instead of the software supporting the organization (e.g.

Davenport, 1998). A large body of research has addressed general implementation and adaptation problems in a wide range of industries (Akkermans & van Helden, 2002; Benders, Hoeken, Batenburg & Schouteten, 2006; Cadili & Whitley, 2005; Dery, Grant, Harley & Wright, 2006; Fleck, Webster & Williams, 1990; Gosain, 2004; Grant & Hall, 2005; Grant, Hall, Wailes & Wright, 2006; Light & Wagner, 2006; Shepherd et al., 2009; Webster, 1991b).

2.4 Interdisciplinary challenges: Planning in socio-technical contexts

The difficulties in production planning and scheduling are amplified through the hybrid nature of function allocation between human planners and computers in enterprise resource planning systems. This development – to a varying degree across different industries – poses interdisciplinary challenges that have only been addressed recently (Cegarra & van Wezel, 2010b; Higgins, 1999; Shalin, 2005; van Wezel & Jorna, 2001).

Quite fundamentally, as Shalin and McCraw have described in detail (Shalin & McCraw, 2003), there is a problem of *representation* when dealing with plans as hypothetical constructs. A plan might co-exist with other plans, it has a history of changes and amendments, and it requires a user that is capable of memorizing all these modifications in order to interpret and use the plan in a meaningful way (Shalin, 2005). As Vicente and Burns (Vicente & Burns, 1996) reported, even in environments of process control which require planning to a lesser extent than manufacturing, operators are not relying on the representations that are at their discretion because they need more information about the status and trustworthiness of the sensor data displays. This furthermore requires what Jamieson and Vicente (2005) have termed 'mode awareness' - an actualized knowledge about the control module state and behavior. Effective control therefore involves more than accurate mental models of the process and feedback, it demands a skillful control practice that requires knowledge about the system that is controlled *and* the controller unit (i.e. the ERP software system in the case of manufacturing).

In order to tackle these problems it is necessary to understand plan features that go beyond mathematically specified criteria like machine utilization or distribution of workload. In his analysis of plan features that can affect re-planning, Kontogiannis (2010a) focused on plan characteristics related to complexity, coupling and control. In a comparable but not identical way, Hoc (2006) argues that plan features regarding de-synchronization, abstraction and anticipation can strongly influence the practicality, execution and adaptation of a plan. Furthermore, various organizational design aspects are influencing interactions between departments to cope with short-term coordination and planning issues (Nauta & Sanders, 2001; Windischer, Grote, Mathier, Meunier Martins & Glardon, 2009). Hereby the increase in complexity and loss of autonomy in a globalized economy (e.g. Hanseth, Ciborra & Braa, 2001; Rammert, 2003a) leading to uniforming efforts of formalization and digitalization of the workplace, combined with a parallel and paradoxical trend towards informal personal networks among professionals, also account for shortcomings of these structuring efforts (Funken & Schulz-Schaeffer, 2008).

In a case study about the implementation of an ERP system in a large international company, Elbanna (2006) has been able to show that even in the case of relatively rigid and contingent technologies like ERP systems, there is room for improvisation. She found that the people responsible for the implementation were constantly working around the plan to get the system running according to the company's needs. Interestingly, the managerial recount of the implementation process later did not reveal these efforts and thus rendered them invisible by pretending that everything went according to the initial implementation plan.

The role of organizational memory and conflicting knowledge was investigated by van Stijn (2006). In her PhD thesis, she applied a 'memory lens' to ERP implementation processes and organizational change. Van Stijn is describing the sustainability / flexibility paradox when it comes to ERP systems: On one side, ERP systems often become 'fossilized' once the most important adaptations have been achieved. During the enactment and initial 'tweaking' of the standardized technology, local rationalities and knowledge become part of the ERP superstructure. On the other side, due to changes in and around the organization, work processes need to remain flexible to guarantee for successful operations. However, most ERP vendors do not encourage modifications and alterations, or demand very high prices to make them possible. Van Stijn proposes to use the 'memory lens' approach to develop new ways of talking and investigating ERP practices. The lens could also serve to consciously develop a corporate culture of addressing conflicting memories or knowledge.

Corporate culture seems an important factor when it comes to ERP implementations. Kwahk and Ahn (2010) are reporting on differences in ERP implementation and use in Korea. Their findings suggest that large companies have a higher chance to successfully implement 'global' ERP systems based on US and European standards, possibly due to a generally higher computer self-efficacy of its employees. Smaller companies are better off choosing local software products that allow for company- and culture-specific adaptations. There study has certain weaknesses, for instance their measure for success is the intention to use the ERP system in the future. One might argue that the intention does not necessarily lead to the actual behavior. On the other had, their sample of ERP users is very young in comparison. Most respondents to their questionnaire were under 34 years old (78%). Since their main influencing variables were attitude to change and computer self-efficacy, the results might therefore not be easily generalizable to other companies or cultures. Nevertheless their work addresses very interesting questions, namely how culture, attitudes and local practices are substantially shaping technology adoption, implementation and use.

There is a whole series of more recent papers that are discussing the cultural assimilating of ERP systems in specific organizations (Cadili & Whitley, 2005; Elbanna, 2008; Wagner, Newell & Piccoli, 2010). Some are rather critical about the real achievements, suggesting that much of the implementation should be understood as 'organizational façades' that are hiding away the actual state of affairs (Abrahamson & Baumard, 2008). Others are showing that in an academic environment, the creation and elaboration of workarounds became a 'technological zone' of its own, hereby providing a space of freedom and resistance to the overwhelming 'governing' principles built into the standard ERP system (Kitto & Higgins, 2010). Discussions revolving around the standardizing or culture-ignoring nature of ERP systems are related to the question of agency in this context. Rose and Jones are proposing a perspective that takes into account both aspects, human as well as machine agency. In their view, machines facilitate and enable some parts of the human exercise of agency, but constrain other parts. Human agency is different from machine agency in terms of purpose and awareness, therefore having the potential to structure this 'double dance of agency' (Rose & Jones, 2005).

Already in the early days of computer algorithms for planning, Sanderson has argued that the usual performance measures will not be able to capture the most important reasons for which humans are needed in such systems. She wrote: "In particular, humans may be needed because they can solve ill-defined problems on an intermittent basis rather than well-defined problems on a regular basis. Their continued ability to do this requires a sound mental model of system properties and constraints and a clear mental picture of the current system state (Sanderson, 1989, p. 662)."

These short excursions into the domain of computer-supported production planning and scheduling were intended to provide a basic understanding of this Human Factors research domain. The next chapter is dedicated to the psychology behind planning and scheduling activities, as well as some critical observations related to cognitive engineering efforts within enterprise resource planning systems to support them.

3 Human Factors in planning and scheduling

> Much scientific research has in fact the same logical character as detection. In a piece of criminal detection, the detective knows that a crime has been committed and some facts about it but he does not know, or at least cannot yet prove, the identity of the criminal (Bhaskar, 1998a, p. 29).

Computer-supported planning and scheduling raises research questions in cognitive as well as organizational ergonomics as I have lined out in the last chapter. Given the importance of these problems for industrial organizations there has been a significant amount of scientific work done related to all theses aspects (reviews can be found in Crawford, 2001; Hoc, Mebarki & Cegarra, 2004; Sanderson, 1989). On an organizational level, the re-integration of a certain degree of planning and scheduling has been one of the goals of socio-technical design approaches since their beginnings in the 1940s (Trist & Bamforth, 1951), and with more vigor during the 1980s and 1990s in Scandinavia and Germany due to an increased interest in the 'humanization of work' (cf. Badham & Schallock, 1991).

Sanderson (1989) provides an engaged and thorough review of research on human planning and scheduling. She discusses the general nature of the scheduling problem, different approaches to automation and laboratory as well as field studies that have been conducted in a period of 25 years. The laboratory studies she reviews have mainly focused on three aspects: (1) comparing unaided humans with normative scheduling rules, (2) studying interactive systems of humans and computers, and (3) studying the effect of predictive and graphical displays on scheduling performance. The review shows that generalizations are difficult, because the tasks that have been studied are only comparable to a certain extent. Moreover, due to the accelerated technological development many studies are outdated or not relevant any longer. Field studies mostly involved highly experienced schedulers with little decision support. Sanderson concludes that more coordinated research efforts within Human Factors is necessary to establish reliable knowledge about the interaction of humans and computers in the field of planning and scheduling. In a first stage, a thorough analysis of the scheduling task domain is needed. For example, some studies found that schedulers spent 80-90% of their time in identifying problem constraints rather than actually sequencing and dispatching job orders. In a second stage, features of the problem and the interface should be manipulated in a systematic way to gain insight into cognitive strategies of schedulers (Sanderson, 1989, p. 663).

Crawford and Wiers (2001) refer to the research that has been reviewed by Sanderson as 'first generation' work, motivated mainly through scientific interest in the task of scheduling, and limited through a somewhat restricted availability of computer systems to support the task. In their own review, they summarize seven theoretical and ten field studies that had been published between 1990 and 2001. They refer to these as the 'second generation' of Human Factors research in scheduling, which - according to the authors - was mainly motivated by industrial needs that had developed by the end of the 1980s in the wake of flexible and lean manufacturing within globalized supply chains. Mass customization and delivery performance became a characteristic of the product itself and demanded more planning and scheduling. In parallel, computer technology became widely available to provide support for PPS tasks. However, there was a lack of applicable theory to be used in the design of software tools dedicated to PPS, which in turn led to the 'second generation' research efforts (Crawford, 2001, p. 26). McKay's doctoral thesis can be regarded as a starting point for this type of research (McKay, 1992b). Wiers (1997) and Crawford (2000) took this approach further in their own doctoral research. According to them, a 'third generation' of research is beginning to emerge that is motivated by the perceived need for an integrative framework that allows for a more coherent Human Factors approach in both academia and industry. Crawford and Wiers (2001, p. 30) mention three shortcomings of previous research in generations one and two: The lack of coordination, the lack of focus, and the lack of building on previous research. They further identify eight topics for Human Factors research in PPS that have not been systematically investigated or addressed up to date (2001, pp. 31-38):

(1) Cognitive issues: Planners' mental representations, their situation awareness and the role of cue recognition are not well covered by research.

(2) Decision making (as a separate cognitive issue): The role of context-specific heuristics, dynamic adaptation of decision strategy and the problem of delayed or unavailable feedback (cf. Wiers, 1996) are not well understood so far.

(3) Environmental factors: Organizational culture and job specification affects scheduling behavior, mainly through the perceived function and use of plans and schedules. The role of the researcher within the complex structure of an organization is not well reflected upon in general.

(4) Domain-free and context-based factors: Incompatible views that persist in the research community, i.e. the effort to formulate normative taxonomies of scheduling to make scheduling situations comparable (cf. Dessouky et al., 1995) and the acknowledgement that schedulers are only performing successfully when they have intimate knowledge of the work domain and context.

(5) Instability and complexity of the environment: The inherent instability of production systems due to external and internal operational uncertainties leads to various problems of complexity reduction to cope with the dynamics of scheduling situations. Coping strategies in relation to various system states (uncertain vs. stable, overloaded vs. underloaded etc.) are not well researched since they require long-term field studies.

(6) Temporal and production constraints: Some jobs might expire or become worthless over time, others might have a 'procedural utility' for the planner, i.e. to build trust. The constraints are so diverse and embedded in a production facility that they tend to become harder for planners and schedulers to keep track and account for appropriately.

(7) Information issues: The role of information is not straight forward in PPS. Information is widely distributed within an organization, some of it hard to access, and other incomplete or erroneous. Planners are using 'information networks' to acquire 'enriched data' (e.g. historical, cultural or personal information relevant to their task).

(8) Scheduling function or role in practice: There is no consensus about the main function of the scheduler working within a scheduling environment. Some found that the main function is to prepare for and deal with breakdowns of plans rather than the generation of plans.

Hoc, Mebarki and Cegarra (2004) provide a more recent review that is concluding with four persistent major research areas for HF research in production planning: (a) The understanding of human strategies and representations in planning and scheduling, (b) development of efficient interfaces, (c) focussing on the problem of cooperation, and (d) be more aware to reactive scheduling in dynamic situations. The efforts made within a European-funded interdisciplinary network of experts[3] from 2004 to 2008 can be seen as a further attempt to formulate a more general framework and common ground to address these issues (Cegarra & van Wezel, 2010b).

The following subchapters are dedicated to the psychology of planning and scheduling, especially to two selected aspects, decision making and collaboration. I chose these two aspects due to their relevance for cognitive engineering efforts in the field. In a dedicated subchapter, I summarize research findings related to planning and scheduling as decision making. Then I discuss work related to collaborative aspects of planning and scheduling in another subchapter. The subsequent discussion of design efforts and failures will then lead to critical observations on human factors research in the domain of manufacturing control.

3.1 Psychology of planning

Psychological aspects of planning and scheduling have been a topic of scholarly work mainly in the fields of individual action regulation, especially in problem solving (Hoc, 1988; Morris & Ward, 2005). In addition to that, it became of interest for the Human Factors community through the development and propagation of advanced manufacturing systems, which used computers to develop plans but were still very much dependent on human input and judgment (cf. Sanderson, 1989). There is a fundamental difference between planning for oneself and planning for an organization. As Resch (1988)

---

[3] Human and Organizational Factors in Planning and Scheduling HOPS (EU-COST action A29)

has defined, cognitive work takes place within a 'factual action field' (i.e. the planning of the planner's own actions) and within a 'reference action field' (i.e. the planned or expected activities on the shop floor). In fact, both kinds of planning have their own tradition of scientific investigation and theorizing, as van Wezel and Jorna (2001) have observed. The complexity of the resulting cognitive work related to planning in an industrial setting demands for a bottom-up, descriptive approach, and therefore it is subject to Human Factors, as well as Organizational Psychology (cf. Jorna, 2006; Strohm, 1996). Probably the most salient difference is the required communicable representation of the problem space as well as a feasible solution in the case of industrial planning. Opportunistic planning (Hayes-Roth & Hayes-Roth, 1979), which often takes place in everyday life, is usually not appropriate. Table 3 is showing these and other differences in an overview (cf. van Wezel & Jorna, 2001).

On a more theoretical level, the distinct cognitive characteristics of planning in complex work environments has been discussed by various scholars in the past decades (Cegarra, 2004; Hoc, 1993; von der Weth, 2001). The main goal of these efforts has been to identify cognitive demands on a more abstract level. The discussions therefore are concerned with issues of complexity, uncertainty, cycle synchronicity, process steadiness, multiple and contradictory objectives (Cegarra, 2008). Others have been developing more formal approaches to time-related issues in ergonomics, such as models of temporal reasoning and the notion of 'temporal error' - meaning an adjustment that is leading to an undesired evolution of the system (Hildebrandt & Harrison, 2004; De Keyser, 1995; De Keyser & Nyssen, 2001; Sougné, Nyssen & de Keyser, 1993; Vandierendonck & De Vooght, 1998).

Table 3: Comparison of planning aspects (adapted from van Wezel & Jorna, 2001, p. 282)

Aspect	Dimension	**Human who plans for himself**	**Human who plans in an organization**
Kind of entity	Alone / group	Alone	Alone and group
	Natural / artificial	Natural	Natural and artificial
Process characteristics	Information processing	Internal	Internal and external
	Representation	Internal: hidden and mental	External: various and coded
	Communication	Internal: hidden and mental	Internal and external: mental and coded
	Modeling	Artificial Intelligence (temporal, case-based reasoning)	Operations Research
	Relation planning, execution, and control	Intertwined; flexible adaptation after unforeseen events	Decoupled; inflexible with respect to adaptation
Domain characteristics	Problem space	Ill-defined	Strive towards well-defined
	Planned entities	Sequence of own activities	Alignment between other's activities, capacity, orders
	Constraints / goal functions	Self-paced; self-imposed; easily revisable	Externally imposed, non-paced and difficult to change

In the case of highly dynamic situations such as in supervisory control, planning further involves synchronization and de-synchronization between different time frames. The most obvious relevant time frames are the one related to the process that is supervised and the one related to the planning activities (Hoc, 2006, p. 73). In the case of supervisory control within other domains than process control, for example discrete parts manufacturing, anesthesia or food preparation in a restaurant kitchen, more interrelated temporal reference frames are involved (Akkerman & van Donk, 2009; Nyssen & Javaux, 1996). Cognitive processes in pre-planning, real-time planning, adjustment of plans, and re-planning are therefore of greatest interest for applied research in this domain (Hoc, 2006; Kontogiannis, 2010a).

3.1.1 Planning and scheduling as decision making

Planning and scheduling is to a substantial degree depending on decision making by experienced planners based on their extensive knowledge of the production facilities and processes (cf. Crawford, 2001; Fleig & Schneider, 1998; McKay et al., 1989). This knowledge is applicable only when a planner is constantly updating his or her mental model of the situation in the production system, as well as the planning and scheduling system that controls production. They are highly familiar with the dynamic processes and actors involved and are devoting a substantial amount of time to the gathering and analysis of information. Expertise, especially causal knowledge about the environment, hereby influences cue selection and therefore learning and decision making (Garcia-Retamero & Hoffrage, 2006). But, as I will elaborate further below, many production environments are lacking straight forward causal relationships, which in turn makes decision making difficult.

As a consequence, during a workday, experienced planners are often deciding according to their 'intuition' within the general flow of action: "(...) actors are immersed in the work context for extended periods; they know by heart the normal flow of activities and the action alternatives available. During familiar situations, therefore, knowledge-based, analytical reasoning and planning is replaced by a simple skill- and rule-based choice among familiar action alternatives, that is, on practice and know-how. When, in such situations, operational decisions are taken, they will not be based on rational situation analysis, only on the information which, in the running context, is necessary to distinguish among the perceived alternatives for action. Separate 'decisions' therefore are difficult to identify and the study of decision making cannot be separated from a simultaneous study of the social context and value system in which it takes place and the dynamic work process it is intended to control (Rasmussen, 1997, p. 187f)." One could add that these decisions are often influenced by organizational culture, constraints, beliefs, habits and so on (Yates & Tschirhart, 2006).

Whether or not a decision maker is using a rational or intuitive strategy, depending on the situation, is therefore not well understood. Basically, a common typology that allows the description and comparative analysis of such situations is missing (cf. Cegarra, 2008). However, such a typology would be required in order to analyze and predict routine decision-making behavior in complex environments. Without it, it is only possible to describe the outcomes (i.e. the phenomenon of a distinguishable decision strategy), without being able to draw causal conclusions about contextual influences on strategy selection. The strategy selection as a mechanism should better be understood as part of the expert's knowledge that he or she has acquired through an extended learning process (Betsch, Haberstroh & Höhle, 2002; Glöckner & Betsch, 2008; Mata, Schooler & Rieskamp, 2007; Rieskamp & Otto, 2006).

Given the general repetitive, knowledge-intensive nature of decision making in PPS, it seems reasonable to turn to research on naturalistic decision making (NDM) to look for a theoretical base and empirical work. However, there are not many NDM studies to be found in the domain of PPS (Crawford, 2000; Gasser et al., 2007). One possible explanation for this could be that NDM, as well as classical decision making (CDM) research, is generally done in somewhat different task environments. Table 4 is summarizing some of the characteristic context and task features that distinguish between the three different kinds of decision making environments. The classical decision making environment typically involves well-defined tasks in a decoupled and controlled environment, i.e. a laboratory. The tasks can be accomplished with none or little training. As opposed to that, the typical environment of naturalistic decision making is dynamic, tasks are ill-defined and require expertise. There is a risk of highly adverse consequences, e.g. the loss of lives and/or expensive equipment.

The case of decision making in planning and scheduling environments has characteristics that are somewhere between these two 'extremes'. The environment is dynamic and coupled by definition, and a lot of expertise is required to make decisions. However, the stakes are not as high as in fire fighting or military combat. Some actions are reversible, others can lead to substantial losses in materials, customer satisfaction and therefore profits. In any case, the consequences are real. But what differentiates PPS decision making from the other two is the frequent absence of feedback that can be linked to the individual decision. This has been described as a 'wicked' type of environment (Hogarth, 2005) - leading to impaired learning. CDM and NDM study environments are generally 'kind' in terms of learning. There is a more or less immediate win or loss, a mission success or failure. This is not the case in PPS. Some consequences are delayed in time, others are cumulative. It is very difficult for the individual planner to adjust his or her decision making according to single decisions and their outcomes. Therefore this kind of decision making is qualitatively different form the other two, and it requires great caution when applying theories or models to it that have been developed within CDM or NDM environments.

Although the presentation of decision problems in PPS is somewhat comparable, i.e. the production of a commodity under consideration of available resources, it remains difficult to generalize from one context to the other. The study of decision making in PPS therefore remains very context-bound, limited to a certain industry or even plant (Crawford, 2001, p. 32).

There are only few studies where NDM methods have been applied to PPS decision making (Crawford, MacCarthy, Wilson & Vernon, 1999; Gasser et al., 2007; McKay & Buzacott, 1995). These studies have shown that decision making in PPS is difficult to investigate. Decision processes are knowledge-intensive, distributed, mutually dependent, hidden, and stretched over long time periods (cf. also Wiers, 1996). Furthermore, as MacCarthy, Wilson and Crawford state, the scheduling task "contains elements of both predictable, sequential behavior and also unpredictable, dynamic and context dependent behavior (MacCarthy & Wilson, 2001, p. 13)."

Table 4: Comparison of context/situation and task features between classical decision making (CDM), naturalistic decision making (NDM), and production planning and scheduling (PPS) environments

Study environment	**CDM** (e.g. finance)	**NDM** (e.g. fire fighting)	**PPS**
Context / situation features			
Routine / experience / discriminative learning (Betsch, 2005; Klein, 1998)	No	Yes	Yes
Other tasks (Klein, 1998)	No	Yes	Yes
Noise, heat, poor sight, disturbances, etc.	Low	High	Medium
'Kind' or 'wicked' learning environment (Hogarth, 2005)	Kind	Kind	Wicked
Supportive for mindful practice (Weick & Sutcliffe, 2001)	Rarely	Sometimes	Mostly
Accountability	Low/Medium	High	Medium
Internal and external dynamics	Low	High	Medium
Social translucence, i.e. visibility of other actors and their actions (Erickson & Kellogg, 2000)	(n.a.)	Medium/High	Low/Medium
Task features			
Complexity, i.e. amount of elements, interrelatedness, dynamic effects, transparency	Low to medium	High	High
Higher order goals, badly defined goals and underspecified courses of action (Klein, 1998)	No	Yes	Yes
Significance ('horizon of consequences')	Low	High	Medium to high
Payoffs	Low	High	Low

Irreversibility of consequences	High	High	Medium
Time pressure or temporal restrictions	High	High	Medium
Insufficient information (incomplete, ambiguous and/or erroneous)	Sometimes	Yes	Sometimes
Inducement of intuition and analysis by task conditions (Hammond, 1988; Hogarth, 2005)	Both	More intuitive	More analytical
Negative affect (Betsch & Haberstroh, 2005, p. 50f)	No	Yes (danger!)	Sometimes
Activation of deliberation goals (Betsch & Haberstroh, 2005)	Yes (test situation!)	Sometimes	Yes (job!)

Facing such difficulties, field study results always have to be considered with caution and one has to bear in mind that these findings are somewhat preliminary. Our own work has shown that planners and schedulers use a variety of decision making strategies, but it is almost impossible to identify influencing factors that would predict the choice of a specific strategy (Gasser, 2010; Gasser et al., 2011). Tacit ('intuitive') and deliberate ('analytical') decision strategies seem to be used in parallel, or at times sequentially. Rule-based decision making seems to be the default strategy, however this also depends on the definition of a 'decision situation': Do we only consider situations with at least two explicitly mentioned and feasible action paths or do we allow for decision situations that are more of type 'do something or do nothing'. Another problem is that of the start and end of a decision episode. Without a valid method for process tracing that is feasible in 'real world' environments, field studies always have to rely on observation and post-hoc interviews ('decision probes'). This leads to various methodological problems, for example the instruction to document and describe a decision (e.g. through interviewing or thinking aloud) is biasing the result in terms of a tendency to report deliberate decision making strategies when asked to verbalize cognitive behavior.

Even when we assume that we have a method at hand to track and document decision behavior in PPS, the problem of performance would remain. How can we measure performance? And what would be the normative benchmark that defines a 'correct' decision in terms of the technique that has been applied to solve the decision problem? - Both issues remain open in the domain of PPS (Dessouky et al., 1995; MacCarthy & Wilson, 2001). They are further complicated through the absence of a single representation of what is 'real'. There are several representations present in any PPS situation: The material reality of the shop floor, its digitized representation in the ERP system and the planner's mental model of the situation (both the on shop floor and in the ERP system). All of them are hard to assess and use as a base for scientific work.

Theoretical frameworks that can deal with these difficulties and therefore might guide field research in PPS are not available up to date. Even worse, it remains unresolved within psychological theorizing which are the conceptual relationships between decision making, problem solving and planning (cf. Jorna, 2006). In some theoretical perspectives, planning is a subset of problem solving (i.e. the search for an optimal solution), in another perspective, problem solving is a subset of planning (i.e. usually, several problems have to be solved to generate a plan). Decision making therefore is seen as an activity that sometimes resembles problem solving and sometimes it can be seen as a cognitive activity on its own. Some have argued that if setting a course of action is planning, then decision making and problem solving consist of many very small planning activities (e.g. prior to the selection and execution of a cognitive strategy).

What furthermore complicates this situation is that decision making in PPS changes its nature due to the development of expertise, it is *evolving* with time. A novice faced with a scheduling problem will start to use 'standard' problem solving strategies to come to a solution (or decision). An expert that is confronted with the same situation will more likely decide between alternatives in a different way. As Norros (1995) describes, experts use conceptual as well as reflective orientations within their work domain. Orientations, in her perspective are contextual ways of organizing cognitive activities based on local experience. Expertise therefore always has a constructive character, and is geographically and socially embedded.

Decision making by experts is hard to describe or model without referral to contextual influences and experience due to a considerate amount of work. If this is neglected, there is a risk of over-simplification and - eventually - destruction of the specific knowledge that the researcher wanted to preserve: "Practice has a logic which is not that of logic, and thus to apply practical logic to logical logic is to run the risk of destroying the logic one wants to describe with the instrument used to describe it (Bourdieu, 1998, p. 82)." In the case of planners' work there is room for reasonable doubt that a rational problem-solving approach to decision making can be applied (Winograd & Flores, 1986, p. 146). In any work context that involves some degree of cooperation and competition, actors are considering other's positions, their knowledge and tactics (an excellent description of expert Poker players' considerations regarding their opponents can be found in Bina, Chen & Milgram, 2008), Planners are constantly considering this kind of 'context' when formulating problems and alternative solutions to them. For example, it is sometimes more important to do the 'right thing' than to 'do it right'. Benz (2007) has termed this kind of utility 'procedural utility' - which is often neglected in managerial as well as psychological decision making research.

In short, there seems to be an imminent 'lack of rationality' in PPS decision making (cf. Trentesaux, Moray & Tahon, 1998, p. 345f) and a characteristic 'wickedness' of flexible discrete manufacturing as a learning environment (cf. Hogarth, 2005; Trentesaux et al., 1998, p. 350). These facts pose challenges for basic psychological assumptions about the structure of work contexts and the adaptation to them by humans. Contrary to cognitive activities found in various domains with complex but 'rational' dynamics, such as fire fighting, a 'lack of rationality' cannot fully be compensated through expert learning (Horn & Masunaga, 2006). And in the absence of meaningful feedback short-term feedback in a 'wicked world', such learning becomes almost impossible. Nevertheless, many psychologists and engineers are still assuming that decision makers are rational in their choice of well-defined alternatives using - however biased - utility functions. But this view has long been contested, with mixed success (e.g. Gigerenzer, 2001; Hollnagel, 2007; Simon, 1956).

Work systems which have an important social component, combined with a lack of transparency and delayed feedback could be called 'wicked environments'. Such environments by definition have a high degree of complexity which makes learning difficult (Hogarth, 2001). Learning from experience and therefore any kind of pattern recognition is impaired. Furthermore, rational analysis is impossible to apply, except in some relatively isolated areas of the work domain. The development of heuristics is limited as well, since the environment - and therefore the decision problems - are constantly changing. If heuristics are 'found', they are functional in a local and relatively constant problem field such as described in the classic study of Dutton and Starbuck (Dutton & Starbuck, 1971). Clearly, a pattern-learning model of expertise is not considered to adequately describe expert knowledge in PPS, since learning in such environments is severely impaired (Hogarth, 2006, 2008; Norros, 1995).

From a systemic perspective every decision has a communicative aspect. Decisions *must* be communicated, otherwise they are not decisions in the narrow sense (Luhmann, 2000). The main function of decisions is their contribution to the reduction of uncertainty in organizations. But they also form, influence and control social structures in organizations, which could be called *intentional* structures as opposed to *functional* structures. Intentional structures provide the overall frame for decision making behavior. Therefore their detailed and formal description contributes to understanding individual and group decision making in organizations (Rasmussen et al., 1994).

More and more, decisions are communicated through collaborative software systems. These formal structures are complemented with informal structures that evolve in parallel (Kleemann & Matuschek, 2008). Formal digitalized systems are a form of intentional structures that influence decision making behavior. Informal structures are often necessary to make the system work, but are they also highly problematic since they are 'invisible efforts' within the organization. Therefore, an insufficient understanding of social processes contributing to the implementation and use of technology often leads to significant and costly follow-up adjustments of enterprise software systems. This is especially the case if some decisions are communicated through the formal structure, whereas others - sometimes conflicting decisions - are communicated through informal paths. Under such circumstances, cooperation and communication can be problematic (cf. Böhle, Bolte, Pfeiffer & Porschen, 2008). To understand these altered work contexts within computer-mediated collaborative systems and to formulate design principles and guidelines is a big challenge for cognitive engineering. It calls for a less 'technicist' and more 'social-cognitive' approach to develop information tools required by today's flexible and distributed industries (Currie, 2009; Orlikowski & Scott, 2008; Rowlands, 2009).

Given the various methodological and theoretical issues in PPS decision making research that persisted over the last three to four decades, it is very unlikely that a commonly agreed upon universal 'human model scheduler' (Sanderson, 1991) can ever be developed and validated. Even a more modest endeavor towards a simple context-based model of human decision behavior in complex planning tasks is hard to accomplish. Except in some very local fields of application, PPS expert systems or algorithm-based decision support tools will be difficult to achieve. The increasingly collaborative nature of planning and scheduling activities - mediated through a software platform - is posing even more obstacles to such an 'engineering approach'.

3.1.2 Planning and scheduling as collaborative activity

Besides the maintenance of a common framework of reference and real-time action coordination, planning and scheduling is an integral part of collaborative activities in general (Hoc, 2001). In an industrial setting, the planning function becomes a specialized task that is allocated to a distinguished team or hierarchical structure within the organization. Within this 'secondary work system' (Wäfler, 2001) planning in itself becomes a collaborative activity.

There is a substantial amount of research on collaboration in planning and scheduling. For example, the Zürich school of sociotechnical analysis and design has addressed collaborative planning within organizations and in between manufacturers (cf. Günter, 2007; Windischer, 2003; Windischer et al., 2009; Zölch, 2001). Others have been concerned with more general questions of routines and flexibility in collaboration within teams or organizations (Feldman & Pentland, 2003; Howard-Grenville, 2005; van Fenema, 2005a; 2005b). Some researchers further address collaboration issues like human-computer function allocation, social visibility and communication of intentions within planning collectives (Rittenbruch, Viller & Mansfield, 2007; Rognin, Salembier & Zouinar, 2000).

A production schedule is a collaboratively established artifact that represents intended future production activities (cf. Figure 6). Planners and schedulers are constantly updating the data on which the calculation of the plan or schedule is based. For example, a planner is interested to create more capacity in a high-demand situation through the reduction of stock replenishment orders that are not urgent. This is achieved through a reduction of safety stock levels. A scheduler might at the same time be interested in expediting orders because of management pressure. Using technical knowledge about production shortcuts he or she reduces the time needed for certain operations under normal circumstances. A second scheduler prioritizes an order through setting the due date very ambitiously in the hope it will be completed on time. The built-in optimization algorithm is omitting all the intentions that lie behind these modifications. The result is an artifact which in many points incomprehensibly deviates from the previous version of the same artifact. In many organizations, the lack of meaning is filled with informal communications *about* the schedule. This sometimes leads to the situation where the schedule is ignored completely by the operational staff on the shop floor.

Coming from a socio-technical and systemic perspective, Debitz (2005) studied characteristics of system design and collaboration structures and individual well-being at work. He found that the transparency of processes is related to perceived stress. A higher process transparency and better feedback reduces stress, but only if it is locally relevant. Furthermore, inter-departmental collaboration is found to profit from knowledge-sharing and learning across occupational boundaries (Bechky, 2003). In the absence of efforts to create shared meanings and interpretations through communication between communities of practice such as engineers, planners and shop floor supervisors, the software-based schedule becomes a more and more detached representation with little or no connection to the reality on the shop floor.

Human Factors in planning and scheduling

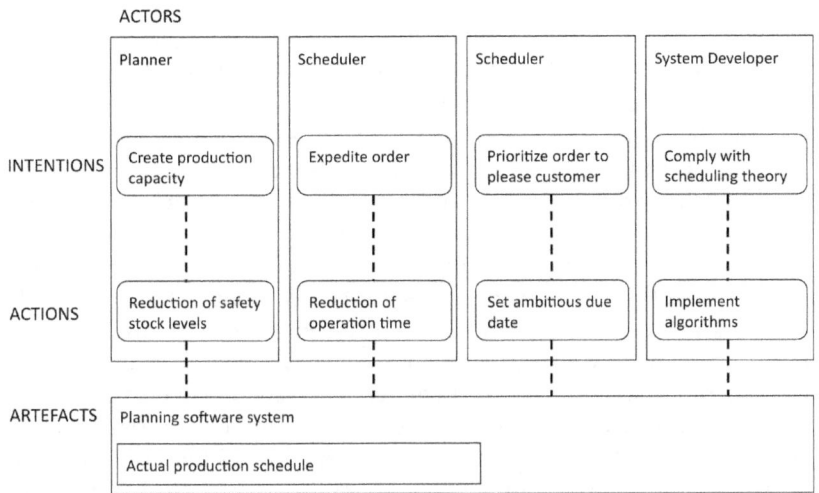

Figure 6: Schematic representation of collaborative production planning

Collaboration is dynamic by nature. Unexpected events demand for negotiations of roles and responsibilities between all involved parties. These processes are necessary to remediate coordination and to institutionalize new structures of collaboration (Clases & Wehner, 2002; Wehner, Clases & Bachmann, 2000). In the domain of planning and scheduling, this is a constantly ongoing process. However, the inflexibility of ERP systems hinders the adaption of new forms of coordination and collaboration in the face of new operational uncertainties. Even worse, their 'black box' nature makes it very difficult to identify the cause of collaborative breakdowns. Accordingly, Christoffersen and Woods (2002) are attributing some problems in human-machine cooperation to a lack of *observability* and *directability* of technical systems. In their view, an increase in automation *without* a parallel increase in coordination structures is bound to create surprises in the form of incidents and accidents. A system which does not allow for the observation of other user's actions, and which impairs knowledge acquisition of other user's intentions is creating a work environment of fear and defensiveness rather than trustful collaboration. In conclusion, ERP systems are a likely source of many organizational problems related to collaboration.

Euerby and Burns (2010) are listing opportunities for sociotechnical design to tackle some of these challenges. They are employing the concept of 'communities of practice' (Wenger, 1999) to describe how complex sociotechnical systems could be made more adaptive. But their support for a concentration on human capabilities, diversity, shared knowledge and participation certainly comes at a cost. However, it seems inevitable to ensure resilient functioning even in highly automated and opaque work environments (Fields, Amaldi & Tassi, 2005). Having said that, it is a far from evident how to establish this kind of 'community of practice' around an 'off-the-shelf' software that is implemented by outsiders of the company or even the industry. In many organizations users of collaborative software tools are lacking problem awareness, are disinterested or reluctant to participate in change processes related to the new technology. This situation could be characterized as 'cognitively ill-defined', where "agents have (...) ill-founded, contradictory and incomplete beliefs about other agents, strategies and rationality (Foss & Lorenzen, 2009, p. 1202)".

To summarize I would like to point out three aspects: (1) Psychological models have so far failed to describe or predict planning and scheduling behavior in organizations. Planning for oneself and planning for others are two different pairs of shoes. Planning and scheduling in organizations demands a substantial amount of domain-specific expertise. (2) Decision making models do not well fit the reality of planners' and schedulers' cognitive work. Their activities are too complex to model in a mathematical sense and at the same time they do not seem to allow for pattern learning and recognition. (3) The collaborative nature of planning and scheduling in a technology-mediated environment creates problems that demand for an interdisciplinary approach that spans from organization science to work psychology and to cognitive engineering.

3.2 Cognitive engineering of enterprise resource planning systems

Production planning software tools that are included in ERP products or installed as an add-on to them can be seen as a form of *automation*. They cyclically update a plan or schedule and are hereby including new or modified production orders. This process is automatic and does generally not allow for human intervention (van Wezel, Cegarra & Hoc, 2011). Planning algorithms are indifferent to intentions and strategic goals of their users. They do not allow for an understanding of the modifications and compromises made by the algorithm compared to its past runs. This creates great difficulties for maintaining situation awareness in PPS. Moreover, despite some efforts to introduce graphical interfaces to PPS software tools (Trentesaux et al., 1998; Upton & Doherty, 2008), the user interfaces of most existing products are predominantly text- and list-based. Due to the enormous amounts of data available, patterns or emergent features are nearly impossible to identify or perceive. Planners are to a great extent depending on electronic communication, phone-calls, coffee-table chats and walk-throughs in production to update and fine-tune their situational mental models (Berglund et al., 2011; Crawford et al., 1999; Karltun & Berglund, 2010).

From a sociotechnical perspective, the fundamental problem of planning software tools in discrete manufacturing is related to the *intentionality* of such work systems as related to their functional integration. Unlike systems that have tightly coupled processes such as power generation or chemical process plants, discrete manufacturing is more loosely coupled or 'intentional' rather than 'functional' - to use Rasmussen and colleagues' basic classification of work systems (Rasmussen et al., 1994). Nevertheless the developers of such tools are assuming smooth and disturbance-free processes that can be modeled within a software with sufficient accuracy. This simplification creates various problems in the actual use of such systems (cf. Trentesaux et al., 1998; Wiers, 1997).

Trentesaux, Moray and Tahon (1998) are putting forward that there is a lack of human factors research in production control. It is generally not well understood how the human planner can be optimally integrated into a complex production management system. Their main focus is on the *responsiveness* of the overall sociotechnical system, since short-term adaptability has increasingly become a competitive advantage in the last decades. A complex management system is needed to control a highly complex manufacturing system. Complexity needs to be reduced locally, however without impeding the overall performance of the production system. Five characteristics of discrete manufacturing systems make this a very challenging endeavor: (1) Their high degree of intentionality, (2) their substantial stochasticity, (3) their reduced tractability, (4) their sometimes chaotic behavior, and (5) their opacity for decision makers (Trentesaux et al., 1998, p. 344f). These characteristics have severe consequences in decision making and cooperation between planners and their production management system, as already discussed further above. Among those, negative dynamics can occur when the management system does not allow for meaningful intervention by human planners and schedulers (cf. Udo & Ebiefung, 1999). Trentesaux Moray and Tahon (1998) are therefore arguing for the development of 'advanced displays' using Ecological Interface Design strategies and methods. Furthermore, they discuss how distributed multi-agent architectures could contribute to the management of stochasticity and chaotic system behavior when combined with such advanced interfaces. However the authors are not very specific on *how* this approach can promote global coherence and compatibility of local decisions. The reason for this might be that it remains unclear how *planning conflicts* are resolved within such a distributed production management structure.

Cognitive engineering modeling approaches have mostly been utilized for highly proceduralized domains such as chemical process plants. According to Rasmussen and colleagues, these work domains could be characterized as predominantly *functional* work domains (Rasmussen, 1986a), where human activity is regulated by the requirements of the system, mostly through standard operating procedures. As Bisantz and Ockerman (2002) describe, cognitive engineering efforts in such domains typically lead to approaches to specify operator information processing goals or quasi-sequential information processing stages which are depending on the actual state of the system. As opposed to that, in work systems that are more *intentional* than functional, like retail businesses, restaurants, or discrete manufacturing industries, activities are constrained through human and organizational intentions rather than laws of nature (like mass and energy conservation). In such systems, "tasks are dynamically assigned to personnel based on available skill level, operator preference and situational demands. The normative task models (...) are not as well suited to focus and analyze observations from field studies in these more flexible situations, in which well-defined, normative procedures are not used to manage tasks (Bisantz & Ockerman, 2002, p. 249)."

Thus, from a Cognitive Engineering point of view, ERP systems and closely connected support structures like advanced planning systems (APS) are posing very challenging design problems. A normative approach seems to be difficult to pursue due to the lack of models and 'norms' (such as laws of nature like energy conservation). A formative approach relies on an incremental design process, which is complicated through the enormous complexity of the technologies involved. In addition to these adversities, the development of ERP systems is mostly technology-driven and users are considered as 'external supporters' that should not interfere except for data entry. However this is in stark contrast with the notion that the success of an artifact depends on the adequacy of the 'user theories' embedded in the artifact (Bisantz & Ockerman, 2002; Kirlik, 1995; Oborski, 2004). Any formative approach is therefore bound to end up in patchwork trying to fix work impairing consequences of a rigid and *prescriptive* technology (Franklin, 1990).

Control capabilities in manufacturing are depending on various organizational and individual resources and structures. Highly standardized 'off-the-shelf' ERP solutions are a risk for overall control capability in a work system. From a control-theoretic point of view, control capability relies on control opportunities, control skills and control motivation (Wäfler et al., 2011). The more prescriptive a technology is, the smaller is the fit to historically grown ways of doing, local expertise and context-specific operational uncertainties that have to be taken into account in daily business. The Contextual Control Model proposed by Hollnagel (2007) also provides an enlightening background for understanding the work-related dynamics in planning and scheduling. Ageing information, delays due to disturbances as well as closing windows of opportunities are demanding various coping strategies for planners. If these are not supported by their electronic planning tools, they risk being disregarded or subverted. Therefore some scholars are proposing context-based approaches to task and structure decomposition as a remedy in the design of PPS software tools (Akkerman & van Donk, 2009) or propose a hierarchical 'mixed initiative' planning approach allowing for these dynamics (van Wezel & Jorna, 2009).

Christoffersen and Woods (2002) are formulating the main problem not as design flaws or human failures, but as breakdowns in *cooperation* between the two sides. In their view, to achieve optimal cooperation between humans and their technology, it is necessary to make machines and algorithms *observable* and *directable,* just like another team player. The key to resolve the problem of observability is 'shared representation' - which includes two basic, but interdependent parts: (1) shared representation of the problem state, and (2) representations of the activities of other agents (Christoffersen & Woods, 2002, p. 4). Or, as Pettersson concludes: "The computer system must be a support to humans' understanding of the environment, in which the computer system is also an integrated part. The goal is to facilitate a consistently shared functional situation awareness (...). This does not imply that both parts must have identical representations, but they must be consistent (Pettersson, 2007, p. 40)." Apart from making an automated system observable it is crucial to provide opportunities for directing it in a desired direction. In many systems, it is by no means optimal to go for all-or-nothing approaches to automation. As Christoffersen and Woods put it: "... we need to preserve the ability of human agents to act in a strategic role, managing the activities of automation in ways that support the overall effectiveness of the joint system (2002, p. 8)."

Within a cooperation-oriented design approach, context-specific, participatory and work-centered design efforts might further strengthen the human-machine cooperation in PPS (Riedel et al., 2011; van Wezel et al., 2011). One core issue hereby is the question of *function allocation* and related to that questions around the adaptability of automated functions. Function allocation principles have mostly been developed for process control applications. In the case of PPS, function allocation has to be reconsidered since the tasks and interactions are not comparable in a straight-forward fashion to process control: (1) Cost is an important factor in PPS. This might explain why interface design usually gets little attention and resources. (2) A complete system failure is unlikely, except for some tightly coupled manufacturing systems. (3) There is a time lapse between plan generation and execution. This allows for the sequencing of planning tasks, which might involve recurrent steps of plan generation. Some errors can furthermore be corrected before they can translate into consequences in production. (4) The algorithm leads to a result which must be approved before it is executed. This requires time which is not available in process control. (5) Failure of automation is not a relevant scenario. Skill degradation is therefore not so much an issue in PPS. (6) Cycle duration from information acquisition to action implementation is generally longer in PPS than in process control. However, the problems that have to be dealt with in PPS can be significantly more complex, in the amount of information as well as the multitude of goals and constraints. The way to approach function allocation in PPS therefore is to decompose the planning and scheduling task into sub-problems and then to look for (cheap, readily available) automated solutions for local problems which might be implemented and used in accordance with the human users' needs and preferences (van Wezel et al., 2011). The question of adaptability is close related to the questions around observability and directability. It remains currently unresolved, since most planning and scheduling algorithms or software tools do not allow for the planner to observe or intervene in the process of plan generation. An *adaptive automation* would provide options to switch between automatic an manual mode for each of the steps or phases of plan generation (Parasuraman & Wickens, 2008).

But, even with adjusted levels of automation and optimal human control over automated functions, the problems around shared representations and observability persists (Dumazeau & Karsenty, 2008). Transparency and social visibility become an evident need of human planners, and related to them are questions of trust and compliance. Making planning and scheduling more observable might even lead to work-related conflicts within the organization, since it would no longer be possible to 'hide' subversive actions within an intransparent and imponderable technology. As a consequence, hitherto unreflected relationships between power and technology might rise to the surface (cf. Clegg & Wilson, 1991). Large-scale information technologies implemented in organizations can be understood as a kind of *institution*, or a contribution towards further institutionalization of the organization. Any sociotechnical design or cognitive engineering practice that neglects this aspect of technology risks failure that might be avoided with a broader perspective (Pettersen, McDonald & Engen, 2010). Unfortunately, there is a lack of theoretical and empirical work on information technology and institutionalization (Currie, 2009).

3.3 Some observations and intermediate conclusions

Given the omnipresence of ERP systems in many industries today and the problems faced by many organizations in implementing and using them, this certainly seems like a relevant domain of application for Human Factors. Especially since there is a persistent lack of human factors considerations in the development of these systems. However, as already indicated in the chapters above, it remains uncertain if the theories and methods that were developed for 'functional' domains like process control or air traffic control can be applied successfully in more 'intentional' domains like manufacturing or services[4].

Related to the question of the applicability of Human Factors to certain work practices are more fundamental issues related to the ontology and epistemology within Human Factors as a discipline. Specifically, the concept of human agency and intentionality is more implicitly than explicitly developed in Psychology and in Human Factors in particular. Moreover, the fundamental philosophical problem of causality is hardly discussed in relation to social structures and technology. Some exceptions are found only on the very edge of the academic community of cognitive engineers and work psychologists. These scholars are trying to overcome the modernist 'concept of humans' (in German: *Menschenbild*) within Human Factors.

But, because these sub-disciplines are historically and culturally *technology-oriented*, humans are often perceived as organisms with limited capabilities which impair a successful adaption to environmental demands (such as those imposed on them by their tools, machines or vehicles). This deficit leads to a desire to intelligently shape and design tools and environments so that these limitations and constraints do not hamper the overall performance of the operator-machine system. It is due to the underlying essentialist concept of technology that one can deduce the *Menschenbild* employed by these disciplines. Humans are mainly understood (and modeled) as consumers/users of technologies, who are at the mercy of this

---

[4] The distinction of 'functional' and 'intentional' work systems has been introduced by Rasmussen, Pejtersen and Goodstein (Rasmussen et al., 1994). In comparison, sociologists of technology have proposed the differentiation of 'material technologies' versus 'social technologies', for example, the classical Weberian buerocracy is a social technology whereas the space shuttle is a material technology (Pinch, 2008).

technology, reacting to it without knowing the broader picture (human as animal rationale as opposed to a more conscious and knowing homo faber - cf. Schlick, Bruder & Luczak, 2010, p. 24f).

Pettersen, McDonald and Engen (2010) are formulating another fundamental critique using Bhaskar's work on social theory (Bhaskar, 1998b). According to them, mainstream Human Factors approaches fail to identify and describe social realities that shape the relationships between social structures and individual agency in organizations. The model building that predominates in Human Factors has a high validity in describing and predicting the "local relationships between people and technology, but support *no* independent criterion for explaining action outside individual (or small group) intentionality (Pettersen et al., 2010, pp. 184, italics added)." Partially, this failure can be attributed to a 'modern', 'linear' and positivist conception of technology, and a disciplinary 'blindness' towards dynamic historical-societal forces that are shaping work systems and technology (cf. Beck, 1992; Hill, 1981; Noble, 1984).

Historically, Sociotechnical Theory (Trist & Murray, 1997; Trist & Bamforth, 1951) is one of the fundaments of Human Factors. It is based on a certain form of *structural functionalism*, the notion that within a society, every entity has evolved through a process of evolution to fulfill a societal function. Through mutual evolution of the social and the technical, the system evolves towards an ideal or optimal design. In this way, Sociotechnical Theory is modernist in its essence, especially also in its optimist embrace of technology. However, especially in its beginnings, it is also critical towards social implications of design efforts. Its proponents believe that negative aspects can be moderated through sociotechnical design, i.e. a joint optimization approach to systems design. A more recent overview that reveals this 'modernist' essence of sociotechnical theory and design can be found for example in the Dutch Integral Organizational Renewal (IOR) approach (van Eijnatten & van der Zwaan, 1998).

In order to overcome limitations due to a naïve modernist perspective on technology, Orlikowski and Gash (1994) have introduced the concept of *technological frames* as an approach to examining and explaining the development, use and change of information technology in organizations. By technological frames, they mean shared cognitive structures concerning technology held by different groups of stakeholders such as technologists, managers or users. Their main objective is to explain difficulties and unanticipated outcomes of technology implementations through incongruences in technological frames. The cognitive structures that they consider as relevant for this endeavor are *assumptions*, *expectations* and *knowledge of technology* collectively held by a group or community. In a comparable way, Ciborra and Hanseth (1998) observe information technology management agendas: They are obvious, sound and look pragmatic. But in reality, they are "deceivingly persuasive (Ciborra & Hanseth, 1998, p. 309)". The believes and expectations held by important groups of stakeholders are of substantial relevance for any discussion of technology. These 'framing' effects also account for scientific work, since the formulation of research questions and the choice of empirical methods depends on them.

Assuming that one could develop a somewhat neutral technological frame about ERP systems, there are still multiple difficulties in applying psychological models to PPS behavior, as shown above. Analytical simplification is leading not only to deception (e.g. planning = decision making), but also serious misunderstandings, be it on the side of the designer or on the side of the user of PPS technology. Reductionistic models lead to 'inscriptions' of behavior that are poorly adapted to the realities of planners and schedulers. It is mainly the intentional and distributed character of PPS activities that challenge these abstract and 'prescriptive' solutions which are bound to be subverted and 'worked around' as a result. The general process of digital *representation* and technological *inscription* - in the case of ERP systems often following a period of applied Business Process Reengineering - can be assumed to be heavily influenced by various stakeholders (cf. Kallinikos, 1995; Zuboff, 1988). The interests of these stakeholders are translated into requirements and standards that are directly shaping ERP tailoring and implementation. In the end, information technology infrastructures are not just networks of data flows and work procedures, they are *sociomaterial* embodiments of actual work practices and institutional arrangements (Ciborra & Hanseth, 1998; Ciborra & Lanzara, 1994; Hanseth & Monteiro, 1997; Orlikowski & Barley, 2001; Orlikowski, 2007).

Analyses and therefore solutions remain insufficient due to a subjectivity-free, uncritical approach and methodology: Naïve descriptions of 'information ecologies' or 'distributed cognition' that are ignoring power structures and ideologies which are shaping social discourses, individual cognitions and artifacts (Dery et al., 2006; Light & Wagner, 2006; Shepherd et al., 2009). They are also ignoring the fact that top-down approaches to technology management sometimes create paradoxical situations due to their lack of systemic 'connectedness' to the environment (Lemon, Craig & Cook, 2010). Or the fact that they tend to be blind to the implementation of political agendas through the use of technology (Córdoba, 2007; Koch & Buhl, 2001).

In addition to these problems regarding representation of real world processes in a networked computer system and their implications for organizations, Ignatiadis and Nandhakumar (2006) have described a even more fundamental dimension of work with enterprise information systems like ERP: The managerial practice or *nature of work* that is afforded by or inscribed within the ERP system. The ERP system's ability to shape user's activities could be theorized as a kind of agency. Whereas human agency is based on intentionality, non-human agency is based on affordance. Referring to a particular software product, SAP, Hanseth, Ciborra and Braa state that it "is more than a pure software package to be tailored to specific needs. It also embeds established ways of using it as well as organizing the implementation project which are further embedded or inscribed into the documentation, existing installations, experience, competence and practices established in and shared by the SAP 'development community', etc. The SAP implementation has been a guiding tool for selecting activities to address, and in which sequence they should be addressed. It has also been a tool and a medium for representing, 'designing', and implementing new work processes (2001, p. 60)."

In a detailed case study, Nandhakumar, Rossi and Talvinen (2005) have been able to show how the process of ERP implementation is shaped both through the members or the organization as well as the technology. They list process shaping influences based on intentionality (human agency), affordance and political/cultural forces. In support of these findings, Rose, Jones and Truex (2005) provide findings from three case studies to discuss the theoretical implications of the notion of distinguishable human and material agencies or influences. In their analysis, neither Structuration Theory nor Actor Network Theory can fully account for the complex interplay of these forces when investigating real technology implementations. Kallinikos (2009) points out that we might even need a more historic perspective to fully understand the scope of the revolution in work practices that comes with the enormous increase in informatization during the past decades. In his view, the ubiquitous digitalization and rendition of reality in today's work environment deserves much more scholarly attention.

These basic assumptions and related problems briefly mentioned above are providing a useful background to understand and formulate existing research issues in PPS. It is critical to attribute the difficulties in PPS research not only to the domain or subject matter, but also to the fundamental shortcomings of a restrictive 'modernist' perspective in Human Factors. More specifically, the interrelatedness of technological structures and individual agency has to be clarified on a theoretical level in order to tackle design problems. To achieve this, it is necessary to discuss the general *modernist* outlook of Human Factors and attempts to overcome its limitations.

The case of production planning and scheduling reveals fundamental issues related to Human Factors theories and methodology. These issues can be traced back to two main problematic aspects of Human Factors as a scientific discipline: An inherent modernist *Menschenbild* and, related to that, an understanding of technology that could be called 'technological essentialism'. Both aspects lead to the same question: How can Human Factors as a discipline renew itself and become more open towards critical and constructivist ideas and approaches?

Sharple and colleagues' approach of 'socio-cognitive engineering' (Sharples et al., 2002) can be used here as an example of a methodology that tries to achieve exactly that. However, it implies that there is one correct or ideal way of working and that design has to optimize the work setting as a whole. In this it is not any different than the sociotechnical thinking of the 1950s: "Socio-cognitive engineering draws on the knowledge of potential users and involves them in the design process. But it is critical of the reliability of user reports and it extends beyond individual users to give an analytic account of the cognitive processes and social interactions, the styles and strategies of working, and the language and patterns of communication, so a to form a composite picture of the human knowledge and activity (Sharples et al., 2002, p. 311)". And further:

"Our aim is to define human-centered systems that are based a sound understanding of how people think, learn, perceive, work and interact (Sharples et al., 2002, p. 311)". These quotes show the modernist undertone that this work carries along also a positivist notion of scientific inquiry. I believe this will not be sufficient if Human Factors really wants to become fit for the challenges that are ahead.

In the following chapters I will sketch out a pragmatic attempt to solve some of these issues within Human Factors as a discipline. It is my intention to develop a theory-conscious methodology or *theory/methods package* to allow for the analysis of work systems that involve complex and powerful technological tools or agents as well as knowledgeable human agents. Hereby I would like to make some restrictions: I do not intend to come forward with a new theory of technology or a new philosophical foundation of Psychology. My perspective is limited, and my propositions are based on my research as a work psychologist in the field of ERP systems. Drawing from this application-oriented focus in my own work I attempt to extend and elaborate the theoretical and methodological base of Human Factors to embrace some of these new technologies, but possibly not all of them (cf. Rammert, 2007, p. 81).

4 A critical theory/methods package for Human Factors

> Analyses, syntheses, insights and intellectual creativity: this is how philosophy as conceptual engineering can help us to design a world in which (...) the marriage of physis and techne may be successful and bear fruit. We need to be stubbornly intellectual (Floridi, 2011, p. 2).

As I have attempted to show in the previous chapters, many problems within the scientific investigation of human factors in production planning and scheduling are related to a modernist/positivist conception of research in the disciplines that are conducting such studies. In order to avoid an unfruitful deadlock between reductionistic modeling approaches and purely hermeneutic efforts to explain reasoning, it is necessary to reflect the theoretical fundaments of both approaches. In order to do so, we have to look into the philosophical assumptions that are implicitly involved, mainly the ontological and epistemological foundations.

As others have argued, both the objective-deductive as well as the hermeneutic-constructive approach are sharing an *essentially positivist* notion of science in general. And it is precisely this categorical distinction of two separate realms of science - often referred to as 'natural' and 'social' science - that can be criticized on philosophical grounds (Archer, 1998). A pragmatic approach would "judge theoretical concepts by their consequences (Rammert, 2003a, p. 291, translation by author)" instead of a judgment based on disciplinary or ideological categorization.

It will no longer be sufficient to let engineers develop technological systems that are then adapted to humans and social structures. Vice versa, it is no longer acceptable that social scientists restrict themselves to the questioning of technological systems. Yet, in order to change these practices, we need to review not only theoretical concepts that are constitutional to these core disciplines, but also their *philosophical foundations*. Only through this excursion into the ontological and epistemological depth of the respective scientific approach it will be possible to come forward with a new, self-critical and constructivist theory/methods package[5].

The fundamental philosophical problem seems to be related to the kind of realism that stands behind specific research positions (Archer, 1998; Bhaskar, 1998b). A *classical realist* position (usually occupied by positivists) employs an objectivist/empiricist epistemology and is advocating quantitative, confirmatory, deductive, laboratory-focused and nomothetic methods. A hermeneutic or *social constructivist* position employs a radically relativist epistemology and is advocating qualitative, explanatory, inductive and context-sensitive methods. Both approaches can be inconsistent in their practice, since they both frequently revert to a *causal realist* theoretical position (Smith, Bennett & Stone, 2006). A way to reduce inconsistencies in social science research is to overcome the duality of (classical) realist versus (classical) constructivist and to restore a *critical* but also *realist* philosophical fundament.

The accelerated development of new technologies in the past two decades has facilitated and enriched the theoretical discussion around critical as well as social constructivist perspectives on technology. These highly networked technologies have particularly made evident that a causal-realist, functional and hierarchical segmentation - for example as described by Hutchins (1995) is no longer the standard of socio-technological systems. Instead, more and more "fragmented and interactive distribution (Rammert, 2007, p. 88, translation by author)" can be found in today's complex sociotechnical ensembles. Moreover, the increased agency of technological systems directly leads to new ways of theorizing about technology (e.g. Bloomfield, Latham & Vurdubakis, 2010; Feenberg, 1999; Latour, 2005; Orlikowski, 2007; Rammert, 2007).

In my understanding, critical or emancipatory thinking essentially requires self-critical reflection on the ontological/epistemological fundaments of science in general and of a specific scientific discipline. It reveals the relativity or historicity of findings within a social, historical and geographical context. It may involve feminist, post-colonial, or other traditions of critical thought. In other words, critical thinking, as somewhat opposed modern

---

[5] Clarke (2005) has called for 'theory/methods packages' – a concept first used by Leigh Star – to be developed and used by the social sciences to make theory-building more transparent.

science, is not blind towards the socially constructed and thus interest-driven nature of scientific thought and practice. It tries to overcome potentially harmful limitations and biases of modernist conceptions of humans, science and technology by making the socially constructed nature of any scientific theory explicit and debatable.

Within this framework of critical thought, I am considering a post-essentialist or social constructivist perspective on technology (cf. Grint & Woolgar, 1997): More specifically a culturally and historically embedded *critical realist* view of technology (cf. Mutch, 2002; 2010). It encompasses the importance of discursive *negotiations* of technology within social structures. Further, it acknowledges "hybrid situations of action (hybride Aktionszusammenhänge; Rammert, 2007, p. 79, translation by author)" and the role of local rationalities/knowledge. Such a perspective puts *agency* and *discourse* dynamics into the center of the study of technology design, implementation and use.

In the following, I will summarize some of the existing social constructivist and critical realist work that can be found in the literature on technology-oriented sciences. This will prepare the following discussion - in the second sub-chapter - on philosophical issues and the necessity for a re-orientation in Human Factors when addressing issues of highly distributed and networked systems. I will discuss three aspects: Firstly, possible reconfigurations from a critical realist perspective within Human Factors as a discipline. Secondly, philosophical assumptions behind some of the analytical approaches in Cognitive Engineering, namely Cognitive Work Analysis. Thirdly, I will conclude with a discussion on ontological and epistemological consequences for Human Factors.

4.1 Critical theoretical perspectives in technology-related sciences

Critical theoretical perspectives understand all knowledge as *situated* knowledge, produced and consumed by particular groups of people, historically and geographically locatable. Claims of universality are considered naive at best and much more commonly as *hegemonic* strategies seeking to silence or erase other perspectives. The core of critical thought is the suspicion that all truth claims are masking or serving particular interests in local, cultural and political struggles (cf. Clarke, 2005, p. xxv). In critical theory there is no such thing as a ideologically 'neutral' place (Knoblauch, 2005, p. 216). In my understanding, a scientific approach can be categorized as critical when it (1) employs a (self-)critical approach to the analysis of its field, (2) has developed a methodology that takes into account power relations and struggles, and (3) is engaging in an interdisciplinary exchange *about* science with sociology, historical and cultural studies or other neighboring disciplines.

When applied to technology-related sciences this would mean that at least a dual approach to technology needs to be taken. The analysis needs to encompass the material elements of technology as *structures with causal tendencies* whose activations are dependent upon *actions of social actors* within a given geo-historical and cultural context (Smith et al., 2006). But, it is not sufficient to merely uncover the distinctive forms through which technology constrains and enables human behavior. In order to follow the critical or emancipatory program, it is necessary to also embed these findings in the discursive network of who is claiming which truth in whose interest.

Having said all this, the question is: Which critical theoretical positions already exist in technology- or work-related studies? How does or should a critical approach to Human Factors look like?

Human Factors Psychology as well as Cognitive Systems Engineering predominantly understand technology as *essence*, as an independent, purposefully designed tool, with human *users* as constrained or incapable of adapting to it. Therefore the underlying research program has been set to study *limitations* of humans and change the design accordingly. Critical or alternative perspectives of humans and technology have not been very influential in Human Factors as a science, but were well received by other design-oriented communities (e.g. Landauer, 1995; Luff, Hindmarsh & Heath, 2000; Suchman, 1987; Theureau, 1992; Winograd & Flores, 1986). Unlike in other interdisciplinary research fields (like Gender Studies or Information Sciences), there is not much discussion or controversy within the scientific community of Human Factors and Ergonomics about what is *human*, and what is *technology* in its essence (cf. Grint & Woolgar, 1997).

Post-essentialist views of technology require a perspective on technology as discursively negotiated *text* (Grint & Woolgar, 1997) or hybrid *actor-networks* including humans *and* artifacts as agents (Latour, 2005). As a consequence, human behavior within such systems is interpretatively dependent on discourse or interaction. In other words, the focus has to be on *agency*, the *habitat* in which agency operates and which it produces in the *course of operation* (Baumann, 1992). In many typical Human Factors domains of application, the technological material or the 'work environment' of the individual actor becomes more and more adaptive, personalizable or modifiable due to an increasing computerization and flexibilization of the workplace. Therefore, to keep in pace with technological developments, it is no longer sufficient to solely base research on relatively stable instances of technology (like plane cockpits or power plant control rooms).

Consequently, there is a need for a thorough discussion of what *is* technology, and how does this conception influence Human Factors as a discipline. The fundamental question hereby is, do we understand technology rather as a given structure that has an essence (i.e. it does what it has been designed for)? Or do we understand technology as something rather historical, interpretative and socially constructed/transformed?

4.2 Contributions to the development of a critical perspective in Human Factors

Drawing on the seminal works of Berger and Luckmann about the social construction of reality (Berger & Luckmann, 1967), several groups of scholars in different countries have began to investigate the social construction of technology (cf. Bijker et al., 1987; Dahlbom & Mathiassen, 1993). Hereby the focus was on the interrelatedness of different actors within technological change, for example scientists, engineers, politicians, consumers and corporations. What unifies these attempts is an effort to avoid ahistorical sociologistic or psychologistic explanations (Berger & Luckmann, 1967, p. 187). Mainly using case studies, these authors have shown that inventions are hardly ever purely scientific, but almost always also driven by considerations related to societal change and commercialization. The research program linking these efforts has also been called Social Construction of Technology (SCOT). Going even further, Latour, Callon and Law proposed to drop the distinction between human and non-human actors and consequently started to use the term 'actor-network' to describe what is happening inside the 'black box' of technology development and diffusion (Actor-Network Theory; e.g. Latour, 1993; Law, 1991).

More recently, in her study on 'centers of coordination', Suchman aptly argues for the inseparability of technology and practice. She finds that there is an "inseparability of technologies and the activities of their use. This includes locating the functionality of technological artifacts not in particular devices, but in densely structured courses of action involving the assembly of heterogeneous devices into a working information system (Suchman, 1997, pp. 44-45)". As opposed to modernist approaches to the evolvement of humans and their technologies, Suchman's approach does not imply an ideal that can be achieved through evolution or conscious design. Rather, this perspective allows for the emergence of certain practices and embraces a much more chaotic and less deliberative view on development. In her studies she found that "people are engaged in a continuous process of making the environment work for the activities at hand. In doing so, they leave the mark of their activities on the environment in ways that set up the conditions for subsequent actions. Along the way, the workspaces, furnishings, technologies, and artifacts are experienced as more and less focal or contextual, negotiable or resistant, enabling or constraining of the work that needs to be done (Suchman, 1997, p. 46)".

Suchman, since the publication of her seminal work on 'Plans and Situated Actions' (Suchman, 1987), emphasizes a perspective on purposeful human action as situated in the *context* of particular circumstances. Her position is a strong and forceful critique of the modernist user modeling and 'planning-based' design approaches widely used in engineering disciplines. In doing so, she draws upon ethnomethodological concepts and theories that have since been very influential in the HCI community (Dourish & Button, 1998). Her contribution could be understood as social constructivist in the sense that she acknowledges the dynamic nature of behavior and technology use in practice. The fact that computers are more and more ubiquitous has put some importance in these theoretical considerations, even though they are not entirely new in Psychology and Social Anthropology (cf. Barker, 1968; Boesch, 1991; Hacker, 1995; Hacker, Volpert & Cranach, 1982; Leontjew, 1982; Lewin, 1930).

In parallel, this understanding of the dynamic nature of individual behavior, the social and physical environment has led scholars and practitioners of Socio-Technical Design (STD) to develop new models and concepts of how humans, organizations and technologies interact in the 'real world', i.e. when they have to deal with uncertainty and complex interactions (e.g. Clegg, 2000; Frese, 1988; Grote, 2004). Challenges herby are mainly related to the complexity of work systems, in terms of their distributedness, the number of elements involved, the lack of transparency and environmental uncertainty.

One of the main questions in Socio-Technical Design is: How to ensure a healthy amount of (local) self-organization within distributed, technologically mediated and tightly coupled systems? Systems theory demands a sufficient amount of autonomy - also called 'requisite variety' - to ensure an adequate answer to or elimination of uncertainty (Ashby, 1957; Luhmann, 2000; Vicente, 1999). This has implications not only in the 'primary work system' concerned with production, but also the 'secondary work system' that is planning, scheduling and controlling the production process (Wäfler, 2001).

In all traditions, SCOT, HCI as well as STD, a static view of knowledge is dysfunctional when shaping and designing technology since knowledge is dynamic within 'communities of practice' (Wenger, 1999). It cannot be separated from the knower (Euerby & Burns, 2010). Therefore 'praxis' cannot be described objectively, it has to be analyzed in relation to social roles, individual understanding and the situatedness of acts within a network of social actors. In order to understand subjective acts, it is crucial to uncover the relations between social structure, technology and praxis (Axel, 2003, p. 39).

Accordingly, in Cognitive Systems Engineering and Information Sciences, two directions have been taken by different streams of research: The first one is committed to a more detailed description of ecological niches formed by technologies and their co-evolution with cognitive strategies and abilities by humans at work in such niches. The second one is aiming at the broader picture of organizational change, work cultures, power relations and local rationalities that govern human behavior and technology use in a particular setting.

4.2.1 Technology as 'ecological niche'

In the past three decades, psychologists have shown that cognition is highly adaptive to problem domains and social contexts. Humans behave 'boundedly' rational, that is adequate to the situation and the resources at hand - in a 'fast and frugal' but efficient way so to speak (e.g. Gigerenzer, 2000, 2002; Gigerenzer & Goldstein, 1996; Gillespie, 1992; March, 1978; Payne, Bettman & Johnson, 1993; Simon, 1956). However, such *heuristic* behaviors are not sufficient for novel situations that demand creative 'work around' solutions in complex work domains. Such situations require an understanding of the work system as a whole, with the individual machine or process in the center. Through a thorough analysis of the ecological niche in question, information requirements for the design of 'smart instruments' can be identified (Vicente & Rasmussen, 1990). Instruments that relate to the human understanding and representation of the problem domain are as such intended to form an 'information ecology' to facilitate problem solution and system control (e.g. Nardi & O'Day, 2000; Vicente, 2002).

More fundamentally, there has been some criticism towards experimental approaches in Cognitive Engineering from a Ecological Psychology point of view that refers to the works of Brunswik and also Gibson. In its essence, it is an engaged discussion about the ecological validity of experimental settings and the generalizability of experimental findings. An example is the debate between Vicente and Payne (Payne & Blackwell, 1997; Vicente, 1997). Such debates on methodology are missing a point in my view, since they merely problematize the methodology of their discipline, hereby only taking into account one aspect of critical thought, the dialectical interdependence of thinking and the contexts of action, but not at all reflecting upon the interrelatedness of their own discipline, its methods, its fields and the societal/political environment around them.

Another example showing some of these shortcomings in Human Factors is the discussion of Cognitive Engineering practice by Marmaras and Nathanael (2005). They state that from the very beginning, Cognitive Engineering was meant to be ecological and that the main goal of the discipline is to change behavior, in order to improve overall system performance, i.e. to avoid errors and accidents. But, interestingly, their description of a cognitive engineer's work is almost completely ahistorical and apolitical. The authors mention one type of 'world-to-study', using a categorization initially proposed by Rasmussen and colleagues, where "the historical memory (...) takes the role of the constraining structure in the form of conventions or norms (*if a structure can be considered at all*) (Marmaras & Nathanael, 2005, pp. 121, italics added)". Whilst there is nothing wrong with this observation as such, it is still very telling in the light of my present discussion.

Even more intriguing, none of these mainstream Cognitive Engineering works refers to the basic dialectical and dynamic nature of human technology use: Technology-mediated human activity is *forming* subjects and *formed* by subjects (e.g. Holzkamp, 1992; Pinch, 2008; Tolman, 2009; Zahavi, 2005). If one considers 'smart instruments' in a power plant control room, this might not be at the core of the engineer's interest. But when we consider a word processor or a specific tool to work with numbers, it becomes evident that these technologies have changed and are still changing the way humans work and think.

Cognitive engineers, I believe can be concluded, merely adapt *strategically* to the uncertainty and dialectical nature of their field. What they miss out is a critical look at their own discipline in terms of ontological assumptions and epistemological consequences. Cognitive Engineering is modernist in its approach, there is a widespread belief that with better analytical tools and refined research approaches the discipline will reach a better understanding of the worlds-to-study and improve its impact on reality. There is only little awareness of the more fundamental problems of Cognitive Engineering that would be formulated from a critical, emancipatory perspective on science and technology.

4.2.2 Organizational change and technology

Whereas Engineering as a discipline always used to be closer to the technological artifact than the human using it, other disciplines in the domain of work and technological change should have a broader view that includes the social context. There are some examples from Industrial Sociology and Information Science that I would like to mention here.

In his seminal work on the introduction of numerical lathe work stations in American manufacturing companies after the Second

World War, Noble (1984) describes how the agenda of managerial politics greatly influenced the development and introduction of this technology and how resistance by workers' unions, and their striving towards decent workplaces was without much success. In his edited book on a 'Sociology of Monsters', Law has collected a range of contributions that address the interplay of technology and power in organizations (Law, 1991). One of these contributions, Webster (1991a) analyses the introduction of computer-aided production management software in industrial organizations. She concludes that when industries implement technologies, they are essentially implementing manufacturing programs. These are overlaid with local structures and manufacturing practices which in turn are articulations of local power relations, patterns of expertise and skill, and managerial prerogatives to organize production. Technologies are "both an expression of practices already in use in industry and a plea for future ones (Webster, 1991b, p. 217)". This is reflected in a more recent collection of case studies by Sumner (2009), where the strategies leading to a successful integration of a new technology, more specifically an ERP system, consist exclusively of measures to ensure the *alignment* of both the new and the old structures and practices.

However, there might also be a lack of theoretical fundaments to account for social dimensions when explaining individual behavior. Pettersen, McDonald and Engen are arguing "that when viewed in relation to theoretical antecedents the social theorizing of safety in socio-technical systems is dominated by cognitive and constructivist based approaches *not* capable of capturing the relationship between social structures and individual agency (Pettersen et al., 2010, pp. 184, italics added)." Concerning the underlying reasons for this shortcoming they state: "There seems to be a reluctance to describe social processes as social processes - they have to be reduced to mental processes (Pettersen et al., 2010, p. 185)". In stark contrast to the observed omittance of the 'social' in engineering practice, Grint and Woolgar (1997) are describing how the design process of a software involves a two-way process of mutual *configuration*. In their view, "what counts as an effect (or even a machine) is taken to be a social process involving the persuasive interpretation of information and the convincing attribution of capacities (Grint & Woolgar, 1997, p. 33)". This radically different perspective puts much more emphasis on communication and persuasion. But here, too, the 'middle layer' of power relations and social influence on technology use is lacking.

Based on traditions of usability-oriented design approaches, Sharples and colleagues (2002) are proposing 'Socio-Cognitive Engineering' as a solution for the systematic inclusion of the social dimension in the development process. However, their notion of what the 'social' is remains very vague. On one hand, it seems that these authors understand the social as 'users' or 'informants' and on the other hand the social seems to refer to the 'organization' as the overall structure or work environment (Sharples et al., 2002, p. 4). In their paper, there is no reference to theories of social structures.

Technological settings can be seen as structures that have certain specific ways of behavior inscribed in them. In this sense, they are *'Gestell'*, in Heidegger's terminology, meaning that they are framing nature - including humans - to turn it into a resource that can be disposed of. These *inscribed behaviors* have been called 'action programs'. A technology is acting in imposing inscribed action programs on human users (cf. Ciborra & Hanseth, 1998, p. 313). If these action programs are dysfunctional, users sometimes develop *anti-programs*, leading to a dynamic negotiation process concerning the design and use of a technology.

This partially explains why Human Factors methodology is facing difficulties in environments like the planning and scheduling of industrial production. The CTA method (Crandall, Klein & Hoffman, 2006) for example neglects not only political and socio-cognitive processes in technology design, but also other important 'silenced' aspects like *awareness* in cooperative systems (Paay, Sterling, Vetere, Howard & Boettcher, 2009; Rittenbruch et al., 2007; Viller & Sommerville, 1999). Such short-sighted analytical approaches often endanger technological innovation in organizations.

Within the Human Factors and Engineering Psychology community in general, there are only limited attempts to broaden the perspective beyond Task Analysis (Jamieson & Miller, 2000; Strauch, 2010). Even the more holistic Cognitive Work Analysis approach (Naikar, 2006; Rasmussen et al., 1994; Vicente, 1999) - by the way criticized by Crandall and colleagues (2006, p. 250) for studying workplace dynamics rather than cognition - does not overcome the technicist bias that comes with a neglect of cultural-historical and social-constructive aspects regarding technology. In an overview of

ERP studies by Shepherd (2009), including a classification of the studies by theoretical approach, ontological perspective, form of explanation, treatment of power, politics and conflict, key assumptions, benefits, and limitations, only half of the studies were addressing the issue of structure and agency. Those who did so employed either SCOT, ANT or Gidden's structuration theory as a theoretical-philosophical background.

While this is just a partial and very limited summary of what has been written and said in terms of organizational change and technological developments, the conclusion that these efforts can hardly be labeled as critical and emancipatory is probably not entirely wrong. Crucially, all these attempts to bridge the gap between the social reality of organizations and the dynamics of technology implementation have their strengths and weaknesses. Some of them are including self-critical reflection and some of them acknowledge the role of power relations. But there is no coherent ontology of the social, nor hardly any reference to sociological postmodern thought about technology. Although, for example, Noble (1984) provides evidence for the primacy of managerial control interests over economical logic and Grint and Woolgar explain how to critically 'read' technologies (1997), there is little at hand for orientation in Human Factors in terms of concepts or theories that could clarify more in detail the actual mechanics of discourse, power, intentions and politics in technology development and implementation.

4.3 Technology in the light of Critical Thought

Drawing on the works of Latour (2005), Grint and Woolgar (1997) describe how powerful societal players are in some cases able to inscribe meaning to technology and even to users. Technology in this perspective is a 'text' that consists of inscribed meanings that are deciphered, interpreted and performed by 'readers'. A technology that is understood as a 'text' can be described like a web of relationships between subjects and objects, with distinguishable but dynamic borders. This encompasses 'knowing' rather than 'knowledge' (Orlikowski, 2002), 'distributedness of activities' and 'collective intentionality' rather than 'tasks' (Engeström, 2004), and 'hybrid activity networks' rather than the 'social' and the 'technical' (Rammert, 2003b). Such a post-essentialist perspective on technology puts human and technical agency in relations to each other. As Rammert (2007, pp. 55-64) proposes, the distinctive characteristics of human agents should hereby not be thrown overboard in an attempt to create symmetry between humans and artifacts, such as in Actor-Network Theory. Furthermore, he argues that besides describing the 'enrolments' of actors in technological systems or networks, the analysis should include the interactions and relations that lead to these enrolments and how they are changed over time (Rammert, 2007, p. 87).

4.3.1 Critical perspectives in Organization Sciences and Information Systems Research

As Luhmann (2000) has observed and very aptly described, organizations are oscillating systems that owe their stability more to a loosely coupled network than to a strictly binding 'technology'. A constant historical flow of organizational decisions that build one on each other, inside or outside formal structures, documented or not is keeping the organization alive in a more or less uncertain environment. The resolution of conflicts and uncertainties within organizational networks has also been described as 'invisible work' within activity networks and information ecologies (Nardi, 1996; Sawchuk, Duarte & Elhammoumi, 2006).

Orlikowski has used Gidden's structuration theory to come up with a theory of technology that takes into account this double nature of technological systems (Orlikowski, 1992). In her paper she develops a 'structuration model of technology' that explains how a dual perspective on technology can be fruitfully put to use in organization studies. She states that: "Technology is the product of human action, while it also assumes structural properties. That is, technology is physically constructed by actors working in a given social context, and technology is socially constructed by actors through the different meanings they attach to it and the various features they emphasize and use. However, it is also the case that once developed and deployed, technology tends to become reified and institutionalized, losing its connection with the human agents that constructed it or gave it meaning, and it appears to be part of the objective, structural properties of the organization (Orlikowski, 1992, p. 408)."

Her model of technology (cf. Figure 7) takes human action as a starting point for technology design and development (a). Vice versa, technology is facilitating and constraining human actions (b). Institutional properties and conditions influence humans in their interaction with technology (c), mainly through professional norms, standards and resource restrictions. But also, the interactions with technology have an influence on institutional properties of the organization (d) through reinforcement or transformation of structures of signification, domination and legitimation (Orlikowski, 1992, p. 415).

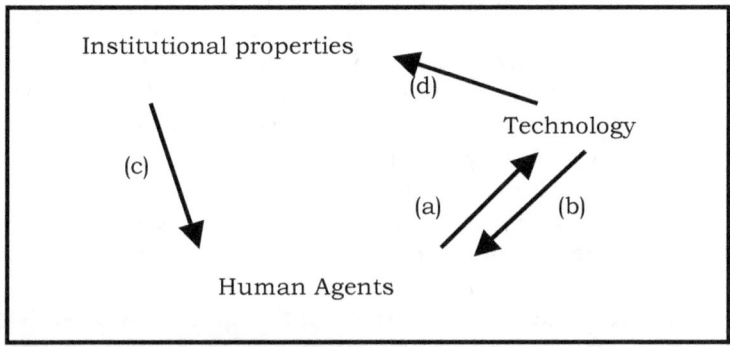

Figure 7: Structurational Model of Technology (adapted from Orlikowski, 1992, p. 415)

The structuration approach to technology has been further developed by Orlikowski in the past two decades. To account for the widely discussed role of emergence and improvisation in technology-oriented scientific disciplines, she proposed a 'practice lens' for studying technology in organizations (2000). In doing so she is at the same time criticizing the notion of 'embodiment' of social structures in technology, which had been advocated by some proponents of the SCOT tradition. Orlikowski firmly sticks to Giddens' formulation that social structures only have a virtual existence, and that they are instantiated in practice. Consequently, she reformulates her theory of technology: "Seen through a practice lens, technology structures are emergent, not embodied (Orlikowski, 2000, p. 407)." In the practice lens perspective, humans are constitutional for technology through their recurrent use of it. Orlikowski calls these enacted structures of technology use *'technologies-in-practice'* and defines them as "the sets of rules and resources that are (re)constituted in peoples' recurrent engagement with the technologies at hand (Orlikowski, 2000, p. 407)". In organizations, technology-in-practice serves essentially as a behavioral and interpretive template for people's situated use of the technology. By means of case studies she is providing evidence that in a given organization various technologies-in-practice may exist for the same technology.

Nevertheless it seems that there is still a lot of work to do in order to understand technology use and to be able to use this knowledge productively. Orlikowski is calling for a more intense collaboration between organization sciences, information sciences and engineering: "Our intent in this essay has been to suggest what might be gained by fostering more interplay between the fields of organization studies and information technology. Our agenda is not to bring about a complete fusion of the two fields, but rather to encourage hybrid research and theory at those points where the two fields intersect. We imagine the hybrid as being different from the mainstream of both fields, possibly in terms of content but certainly in terms of epistemology. In particular, we advocate for research that requires substantive expertise in both technology and the social dynamics of organizing and that embraces the importance of simultaneously understanding the role of human agency as embedded in institutional contexts as well as the constraints and affordances of technologies as material systems (Orlikowski & Barley, 2001, p. 158)."

Following this call, Rose and Jones (2005) have criticized not only the structuration approach with its focus on human agency, but also ANT as a relatively static or descriptive approach to technology. In their theory they propose that both humans and machines have agency (hence the title of their paper 'The Double Dance of Agency: A Socio-Theoretic Account of How Machines and Humans Interact'), but are nevertheless to be distinguished: "(...) humans and machines can both be understood to demonstrate agency, in the sense of performing actions that have consequences, but the character of that agency should not be understood as equivalent. Human agents have purposes and forms of awareness and that machines do not. The two kinds of agency are not separate, but intertwined, and their consequences emergent. Those consequences are also the subject of human interpretations which provide part of the context for future actions (Rose & Jones, 2005, p. 27)." Their main criticism is about the structuration theorist's denial of any agency or causal power embodied in technology. But they also criticize ANT for its symmetry principle. They conclude that a compromise between the two approaches is needed. "The metaphor of the 'double dance' attempts to encapsulate both the intertwined nature of the interaction of human and machine agency, and its part structured, part improvised emergent character (Rose & Jones, 2005, p. 33)."

Interestingly for my own discussion, the empirical background of Rose and Jones is in ERP systems, whereas Orlikowski's field work mainly has been with group collaboration systems like Lotus Notes. There is a big difference in the potential for 'accommodation' or even 'improvisation' of these systems by their users, and this might explain partially why Orlikowski favors structuration theory, whereas Rose and Jones are very sceptical about the amount of agency that can be attributed to users alone.

In Orlikowski's (2005) answer to Rose and Jones she acknowledges that some if not most of their critique - mainly towards ANT notably - is useful and thought-provoking. But instead of using 'agency' when speaking of the causal powers of material things or technologies, she prefers to use the term 'material performativity' (Orlikowski, 2005, p. 185). In a more recent paper she even revised her position on the issue of agency, without mentioning structuration theory nor Giddens one single time. Instead, she advocates the use of the term 'constitutive entanglement' or also 'sociomaterial assemblage' (nota bene: Giddens used the term 'mutually constitutive' to describe the relationship between social structures and human subjects). I quote: "A position of constitutive entanglement does not privilege either humans or technology (in one-way interactions), nor does it link them through a form of mutual reciprocation (in two-way interactions). Instead, the social and the material are considered to be inextricably related — there is no social that is not also material, and no material that is not also social (Orlikowski, 2007, p. 1437)." In a report on studies of organizations that involve technology, Orlikowski and Scott (2008) found that there still is very little research around that tackles this 'entanglement' of social practice and technology. Consequently, they promote a research program on 'sociomateriality'.

Orlikowski's call for a stronger integration of social theory and information sciences has not yet found many adherents. I believe that this is due to the missing theoretical link between social theory and individual behavior, i.e. technology use. Some fundamental questions around social structures, mainly concerning their causal power, are yet to be solved. On the other side, quite contrary to Orlikowski's assumptions, there is not much knowledge about social theories, especially also critical theories, on the side of the engineers and designers of technology. Furthermore, there are very few references to critical thought or emancipatory theories to be heard in organization sciences. In one of these rare exceptions, Kannabiran and Graves Petersen (2010) advocate to consider the 'politics at the interface': The need for an integrated view on design and politics, i.e. "to call to attention the need for a new set of methods, attitudes and approaches, along with the existing, to discuss, analyze and reflect upon the politics at the interface (Kannabiran & Graves Petersen, 2010, p. 695)." Such a set of methods would indeed add a critical edge to 'sociomateriality', which is mainly based on structuration theory. But to achieve that, more theoretical groundwork needs to be done.

4.3.2 Critical Psychology, Activity Theory and Workplace Studies

The specific characteristics of human agency in a technicized and mediatized society have been a concern to some contemporary psychologists (Gergen, 1992; Kvale, 1992; Parker, 2005b; 2007), especially also in the school of Critical Psychology (Fox et al., 2009; Teo, 2008; Tolman, 1994a). The goal of these efforts is to "put that subjective component back into the equation (Parker, 2007, p. 208)". This requires a different approach to psychological research, *involving* the subject as a genuine *co-investigator* instead of excluding it through statistical methods (Tolman, 1994b, pp. 127-144). Such an approach has potential implications for Human Factors and Cognitive Engineering that go beyond attempts of specifying 'joint cognitive systems', and employing 'user-centered' or 'ecological interface design' (Burns & Hajdukiewicz, 2004; Endsley, Bolté & Jones, 2003; Woods & Hollnagel, 2006).

Relationships with technology have been explored in Activity Theory and Interaction Design (Kaptelinin & Nardi, 2006). These theoretical approaches subscribe to a historical and systemic perspective of human agents in organizations, but they strongly focus on a pragmatic and 'apolitical' description of the processes that are evolving or emerging. As Jones (2009) has criticized, there is a lack of political-economical analysis in activity theoretical work about and within organizations. From a purely systemic point of view, the features and characteristics embedded in technologies become less important, and the conditions and forces involved in their design are left out of the equation.

In a more general overview of Psychology as *Wissenschaft*, Valsiner (2009) argues that postmodern *deconstructivism* has failed to renew Psychology as a discipline. But there is hope, since there are many (international) initiatives and ideas on how to overcome the major oversights in Psychology: The re-introduction of dynamic flow, hierarchical order and context through adapted scientific methods. Overall, this could be understood as a program to - theoretically and epistemologically - develop a kind of critical Psychology. However, Valsiner does not mention one important ontological gap that still needs to be addressed by psychologists as well as sociologists: The missing link between social structures and individual behavior. In other words, how do social structures become causally effective when it comes to individual behavior?

The social and material embeddedness of human action have been simplified and de-contextualized in the name of science by mainstream psychologists. Through the 'elimination' of the subject and its replacement by the 'research participant', mainly the historical dimension, or the 'dialectic' between the individual as agent and the social as structure, was lost out of sight. This has led to a situation where applied psychology such as Human Factors finds itself in a difficult position. As soon as the field of study gets complex, an ontological void makes analysis almost impossible. The researcher or analyst suddenly stands in front of a half-built bridge into nowhere land. Or, as Pettersen, McDonald and Engen put it "to bridge the gap between available social theory and applied analysis of socio-technical systems (...) we need to apply social theory that address more explicitly the social dependencies and constraints operating in socio-technical system inde-pendently of any one actor's conceptions of that system (Pettersen et al., 2010, p. 186)." However, this cannot be achieved within a socio-cognitivist perspective, since the challenges are identified "not as problems of just fact or empirical methodology but problems of engaging in ontological formulations and critique (Pettersen et al., 2010, p. 190)." -

One approach that has attempted to shed light into this unknown territory has formed under the label of 'workplace studies' (Button & Sharrock, 2009; Heath, Luff & Knoblauch, 2004; Knoblauch & Heath, 1999; Luff et al., 2000; Rawls, 2008; Theureau, 2000). The scholars using this approach have different backgrounds but share their conviction that a close look into the everyday work practice using mainly ethnological methods (ethnography, ethnomethodology and conversation analysis) is providing rich material. The collection of studies by Luff, Hindmarsh and Heath (2000) show the broad spectrum of approaches and goals within workplace studies. What seems common to these studies is a descriptive approach without any reference to a social ontology or a theory of technology. Technology, in these studies, appears as a condition for human activity, something that is always already in place. Such a position could be interpreted as some kind of technological determinism. This 'static' approach that is focussing on 'thick descriptions' of work practices at a given point in time with a given technological environment is criticized by Engeström in the same book. Engeström stresses the need for a developmental perspective and proposes more interventionist methods to account for the dynamics of change in work environments. But Engeström's approach as such has been criticized in turn as being apolitical. Alvis (2007), while acknowledging Engeström's theoretical contributions to the field, criticizes his empirical approach. He finds that "although contradiction is a central category for Engeström, in its application, notions of social antagonism, exploitation and oppression become side-lined; this in part derives from an interest in the development of the collective worker in the activity system. In addition, contradiction is accented towards tensions or, in his terms, disturbance(s), within an activity system that lead to transformation or some sort of adaptive change (Avis, 2007, p. 165)".

In sum, descriptive studies of technology, such as SCOT or ANT, as well as empirical work using Critical Psychology, Activity Theory or Ethnography cannot account for the local political forces and subtle influences during technology development and use, but they can provide evidence for the existence of such forces in actual technologies-in-use. Their weakness is a non-existent coherent theory of technology as well as analytical tools that are sensitive to power-inflicted contradictions and potentials for resistance in technology implementations.

A critical theory/methods package for Human Factors 85

As I have already mentioned in chapter 3.3 it is not my intention to reinvent the philosophical grounds of Psychology or come up with a new theory of technology. Both projects would by far exceed my possibilities. My interest is in the particular case of ERP systems, their characteristics, their consequences and implications for Human Factors as a discipline. I conclude, referring to the literature cited above, that there is a need to step back one or two steps and to reconsider the modernist base of our discipline, in order to advance into more complex fields such as ERP. Especially, two theoretical building sites seem to be essential to achieve that:

1. The ontological base of social structures as influential entities for technology development and use.
2. The epistemological base of a critical, self-reflective methodology that includes all aspects of technology-in-use without omitting historical or political dimensions.

4.4 Philosophical re-orientation

So what *are* social structures? Do they exist and if so, in what forms? For example, in their book on humans and technology ('Acting with Technology'), Kaptelinin and Nardi (Kaptelinin & Nardi, 2006) are proposing a modification to ANT, namely a typology of agencies to distinguish between human and non-human actors. Their proposition could also be understood as a kind of ontology of agency. However, the philosophical base of their classification scheme (Kaptelinin & Nardi, 2006, p. 244) remains weak: What exactly are 'social entities' and what is their common characteristic? What role are discourses and ideologies playing? How would they argue against the notion that 'social entities' (or structures) do *not* exist independently (for example by Giddens)?

Starting from descriptive or pragmatic approaches to technology and human agency it is hard to counter the argument that both are 'mutually constitutive' (Orlikowski, 2007), and such a thing as social structures do not exist. But if we subscribe to the insight that there is no such thing as ideology-free science, we need to come up with a position that unveils the political background (consciously or unconsciously) promoted by such a stance. In my understanding, to say that technology and humans are 'mutually constitutive' is telling only half the story. It remains a 'modern' approach in the sense that there is a hidden technological determinism behind it. The implicit assumption is that there will be some kind of negotiation and resolution, but value-free. After some rounds of structuration, the best-functioning, fittest systems will survive. This might work for some kind of software applications or gadgets, but can this really be the last word when it comes to technology of all kinds? If science *is* political, then technology is also political. Thus, from the point of view of a critical academic, the second part of the story must be told as well (for more elaborate reflections on the 'missings' of social constructivism and ANT see Radder, 1992).

The only way to overcome the Structuration Theory approach is to tackle its assumptions. To start with, we can argue that humans always find themselves thrown into a society that pre-exists. Social structures are already in place, enacted or not. Therefore, what is needed is some kind of *realism*, a philosophical position that assumes the existence of things beyond the 'actual'. According to Collier (1994, p. 6) a *realist* theory makes the following knowledge claims:

- A claim of objectivity, i.e. what is known would be real whether or not it were known;
- A claim of faillability, i.e. claims can always be refuted by new information;
- A claim of transphenomenality, i.e. knowledge can relate to phenomena beyond appearances, to underlying structures that make appearances possible. Social systems and DNA are two examples of transphenomenal knowledge;
- A claim of counter-phenomenality, i.e. knowledge can also contradict appearances.

Critical Realism (CR), as a realist philosophy of science, claims that the social world cannot be theorized or explained successfully without paying explicit attention to its ontological foundations (Elder-Vass, 2010, p. 69). Critical realism (Archer, 1998; Bhaskar, 1975; 1998b; Lopéz & Potter, 2001; Mutch, 2002) is based on an ontological realism, in the sense that it assumes a reality of being that is intransitive, transfactual and stratified. It encompasses an epistemological relativism that asserts the relativity of all knowledge, which it considers as transitive in nature. Nevertheless it argues for a judgmental rationality that is able to distinguish between more or less accurate theories about the real (Archer, 1998).

The difference between Critical Realism and Structuration Theory, apart from the ontological orientation, is that Bhaskar's transformational model of social activity (cf. Figure 8, Bhaskar, 1979, p. 36) allows for a temporal order of structural conditioning, social interaction and structural elaboration, without the 'confluence' of social structure and agency. Social reality therefore can be differentiated into analytically discrete moments. This provides social realism with a method that can be used to generate practical social theories within a particular domain. The separability of structure and agency is the predicate for examining the interface between them, upon which practical social theorizing depends (Archer, 1998, p. 202). Nevertheless, it is a fact that social structures only exist in a collective of agents who possess knowledge about their activities. This knowledge itself is a social product and as such object to transformation. Social structures can be said to be real but only relatively enduring and autonomous (Bhaskar, 1979, p. 38).

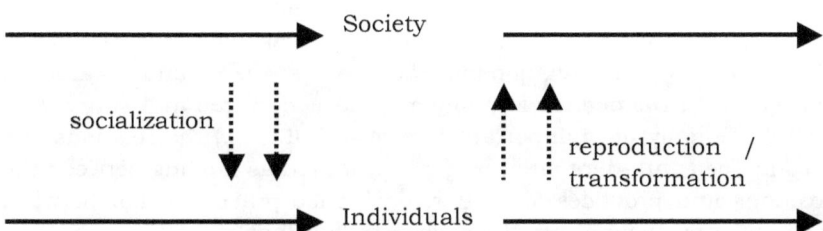

Figure 8: The Transformational Model of the Society/Person Connection (adapted from Bhaskar, 1979)

Bhaskar also speaks of the 'duality of praxis' to describe this ongoing process of socialization and transformation. He proposes to use a set of mediating concepts to designate the 'points of contact' between social structures and individuals: "... social structures (a) be continually reproduced (or transformed) and (b) exist only in virtue of, and are exercised only in, human agency (in short, that they require active 'functionaries'). Combining these desiderata, it is evident that we need a system of mediating concepts, encompassing both aspects of he duality of praxis, designating the 'slots', as it were, in the social structure into which active subjects must slip in order to reproduce it;

that is, a system of concepts designating the 'point of contact' between human agency and social structures. Such a point, linking action to structure, must both endure and be immediately occupied by individuals. It is clear that the mediating system we need is that of the positions (places, functions, rules, tasks, duties, rights, etc.) occupied (filled, assumed, enacted, etc.) by individuals, and of the practices (activities, etc.) in which, in virtue of their occupancy of these positions (and vice versa), they engage. I shall call this mediating system the position-practice system. Now such positions and practices, if they are to be individuated at all, can only be done so relationally (Bhaskar, 1979, p. 40)."

In order to analyze position-practice systems, as 'mediating systems', we need to address mainly the relationships between such position-practices, since they are the fundament for 'generative mechanisms' that produce actual social events. Relations can differ in nature: "In social life only relations endure. Note also that such relations include relationships between people and nature and social products (such as machines and firms), as well as interpersonal ones. And such relations include, but do not all consist in, 'interactions'. Thus contrast the relationship between speaker and hearer in dialogue with the deontic relationship between citizen and state." And further: "Finally, it is important to stress that... (...) the relations one is concerned with here must be conceptualized as holding between the positions and practices (or better, positioned-practices), not between the individuals who occupy/engage in them (Bhaskar, 1979, p. 41)."

An important aspect of critical realist thought is the stratification of reality (Table 5). There are methodological implications for the analysis of position-practices and their relationships related to this differentiation.

Behind the observable empirical world are mechanisms that contribute to the actualization of certain events, but not necessarily. Social structures as position-practices within a network of relations are a kind of such 'generative mechanisms'. Such mechanisms combine to generate the flux of phenomena that constitute the actual states and happenings of the world. They may be said to be real, though it is rarely that they are actually manifest and rarer still that they are empirically identified by humans. Bhaskar distinguishes very decidedly between "the genesis of human actions, lying in the reasons, intentions and plans of people, (...) and the structures governing the reproduction and transformation of social activities (...) hence between the domains of the psychological and the social sciences. (...). It should be noted that engagement in a social activity is itself a conscious human action which may, in general, be described either in terms of the agent's reason for engaging in it or in terms of its social function or role (Bhaskar, 1979, p. 35)".

Table 5: Overlapping domains of the real (Bhaskar, 1975, p. 56)

	Domain of Real	Domain of Actual	Domain of Empirical
Mechanisms	√	-	-
Events	√	√	-
Experiences	√	√	√

In terms of scientific work within such a philosophical framework, one has to part with most of the existing research practices and employ a combined approach between descriptive and hermeneutic analysis. Bhaskar and Archer prefer not to speak of 'induction', they use the term 'retrodiction' for their methodology: "Thus theoretical explanation proceeds by description of significant features, retrodiction of possible causes, elimination of alternatives and identification of the generative mechanism or causal structure at work (which now becomes the new phenomenon to explain); applied explanation by resolution of a complex event into its components, theoretical redescription of these components, retrodiction to possible antecedents of the components and elimination of alternative causes (Archer, 1998, p. xvii)."

The difficulties of the identification of generative mechanisms within social structures (position-practices) are manifold: The inherent openness of social systems, the transitivity of social structures that may or may not be actualized, the diversity of structures and mechanisms. Where to begin?

Hedström and Swedberg (Hedström & Swedberg, 1998) propose three basic types of generative mechanisms:
1. Situational mechanisms (structures influencing agency)
2. Action-formation mechanisms (actual events influencing behavior)

3. Transformational mechanisms (behavior influencing structures)

Elder-Vass (2010) is advocating that social structure should be understood as the causal powers of *social groups*. In doing so, he focuses on a special type of social groups, which he calls 'norm circles'. These norm circles are forming a type of social structure that could also be called 'normative social institutions'. His theoretical analysis of such institutions is intended as an ontological building block that may be combined with others to construct explanations of actual social events.

What does all this have to do with Human Factors? - I believe that there are a lot of benefits for our discipline in this philosophical discussion. Most importantly, it provides a sound theoretical base for the development of an explanatory methodology for our field. It allows us to formulate a perspective on technology as a kind of social structure that entails certain powers (generative mechanisms or *tendencies*). Further, it clearly separates the 'social' from the 'psychological' but without omitting the dynamic influences between them. It clarifies which kind of explanatory statements can be formulated within practical social theories of technology development and use. And - most importantly - it helps us to specify the requirements for the selection of analytical tools. Our methods should not contradict the following statements based on Critical Realist thought:

1. Reality is deep and stratified: The *real* actualizes itself, therefore becoming the *actual*, which in turn may or may not be perceivable or measurable as the *empirical* aspect of reality.
2. Reality exists independently of our knowledge of it. Laws of nature (or *transfactual mechanisms*) can - but not necessarily have to - operate in actual situations. Theoretical explanations are applied for practical explanations of phenomena that they produce in open systems (Archer, 1998, p. xii).
3. Laws cannot be interpreted as conjunctions of events but must be analyzed as *tendencies* of things (Bhaskar, 1975, p. 184).
4. Real structures and forces *actualize* themselves through tendencies to produce certain effects.

5. The *generative mechanisms* that are responsible for this actualization are sometimes interfering with each other, making it difficult to detect causality.
6. Knowledge about the actual (when inferred from the empirical) or the real (hypothesized from the actual) is *transitory*, since truth cannot be proven beyond doubt, only falsified. Critical realist science therefore has to be *epistemologically relativist.*
7. Technology is a socially constructed artifact that has characteristics of a *text*, but also represents a choice of real forces and mechanisms that *determine* the tendencies inscribed into a technology.
8. Analytically, the *geo-historical context* of a phenomenon as well as the specific design of a technology have to be distinguished from the potential that lies behind the actual instantiation, or - in other words - the real that is negated, excluded or submitted in the process of actualization.
9. Empirically, the *morphogenetic process* of institutionalization has to be 'reverse engineered' to understand the actual phenomenon and to frame behavioral data.
10. Theories are exclusively *explanatory* when developed for open social systems (Bhaskar, 1998b, p. 21). Such explanations have to be *practically adequate*, as knowledge they are by definition *practice-dependent.*

In fulfilling these criteria, a Critical Realist method goes beyond a mere functional analysis: "Critical realist evaluation moves from the basic evaluative question - *what* works - to *what is it about this implementation that works for whom in what circumstances* (Dobson, Myles & Jackson, 2007, p. 142)." Such a research approach should be theory-driven, aiming at the modeling of causes as explanations rather than hypothesis testing for generalizations. The results of such efforts are always fallible and subject to subsequent modification (Pearce & Woodiwiss, 2001, p. 52).

In this way, Critical Realism helps to identify potentials for resistance and social transformation that would otherwise go unnoticed. For our discussion of the case of ERP implementations this has far-reaching implications, and a potential to overcome analytical weaknesses of conventional approaches to the study of such technological systems.

4.4.1 Critical Realism, organizations and technology

Critical Realism is not entirely alien to technology- nor to organization-oriented sciences. Dobson introduced Critical Realism to the information systems field and calls for empirical work (Dobson, 2001). Some years later Dobson, Myles and Jackson (2007) published an article with their research findings on an enterprise information system to measure performance. Their conclusion is that "perhaps the greatest benefit of adopting a critical realist underlabouring is the emphasis on deep understandings and context (Dobson et al., 2007, p. 150)". Smith (2006) applied critical realist thought to information systems research. He argues that a critical realist perspective, especially the notion of *tendencies*, can dissolve inconsistencies between theoretical and practical findings in technological determinism and social construction of technology. In his perspective, both approaches contribute substantially to an understanding of technology and human agency. However, a-contextual causal theories of technology are needed to uncover how and why generative mechanisms are triggered (Smith et al., 2006, p. 206).

Volkoff, Strong and Elmes (2007) use critical realism to build up a model of organizational change. In a long-term case study they have used grounded theory methodology to document and analyze organizational change and technology implementation. They found that critical realist ontology helped them to explain the processes that they had observed, something which according to them neither ANT nor Structuration Theory were able to deliver.

In an application of a critical realist methodology (Archer's morphogenetic approach) to the use of information and communication technology in organization, Mutch (2010) sees three gains: (1) Greater clarity about the material properties of technology, (2) links to broader structural conditions arising from the conceptualization of the relationship between agency and structure, and (3) the potential to explore the importance of reflexivity in contemporary organizations, especially in conditions of the widespread use of information and communication technology.

As briefly mentioned in this sub-chapter, there have been some applications of critical realist thought in information and organization sciences. To my knowledge, there has not been any application in the Human Factors domain up to date. Although, in my understanding of Critical Realism fits quite well to some analytical frameworks of

Human Factors, especially Cognitive Work Analysis (CWA) with it's 'formative' approach to analysis and design (Rasmussen et al., 1994, p. 6ff; Vicente, 1999, p. 110). At a glance, its underlying ontology and epistemology do not seem to contradict critical realist assumptions. But since cognitive engineers are rarely explicit in their philosophical background, this issue deserves a closer look.

4.4.2 Cognitive Work Analysis: A realist framework?

Does Cognitive Work Analysis – philosophically speaking - follow a critical realist approach? What are the basic or general ontological assumptions of CWA? Could the framework or its theoretical backing be modified or adapted to fit to a critical realist philosophy? Is there a potential of such a modification for the analysis of 'intentional work domains' like ERP systems?

The initial idea behind the CWA framework was to develop a classification scheme for the description of complex, flexible and highly adaptable work environments. As Rasmussen, Pejtersen and Schmidt argued, "no one exclusive, hierarchical and 'objective' classification scheme will serve to unravel this complexity. What we need is a kind of *teleological taxonomy* (i.e. a pragmatic, goal directed taxonomy useful for the analyst), derived from our need for a framework which can serve prediction of changes in behavior in response to introduction of new information systems. What we can hope to develop is a theoretical framework for *description*, which can also be useful for *prediction*; a framework which necessarily will have the nature of a multi-dimensional, multi-facetted network of interrelated concepts (1990, pp. 17, italics in original text)".

In its taxonomy that implicitly includes some ontological assumptions, CWA postulates 'causal structures' that are constraining human action in work systems. These fundamental structures that are identified through the method of work domain analysis (Rasmussen et al., 1994; Vicente, 1999). At the same time it also assumes 'intentional structures' (identified trough 'organizational modeling') that are also constraining individual action. These structures exist, they are preconditions for human action. In this sense, CWA is indeed a *realist* approach. However, CWA does not formally define the nature of intentional structures nor does it refer to any theoretical work in the domain of structure and agency. What do the developers of CWA mean by 'intentional structures' or 'intentionality of work systems'? Are they referring to agency or social structures?

Rather obviously, an actor, an operator or supervisor, has (conscious) intentions concerning his immediate field of activity, depending on the goals he or she has in mind in a given moment of time. That mostly applies for situations in which a ready-made solution or action path is not available (i.e. in knowledge-based behavior). In such cases, intentions are the result of an analytic and anticipatory thought process:

> "Supervisory control decisions based on functional analysis, i.e., control decisions in unfamiliar situations for which 'know-how' short cuts are not feasible, depend on a prediction of the responses of the system to the intended control action (Rasmussen, 1986b, p. 13)."

In other words, in unfamiliar situations, a human operator is required to perceive the system state, diagnose and interpret that state on the background of his knowledge, then develop a control action (plan) that is in accordance with the goals and objectives of the system (value creation, safety, sustainability), under consideration of the projected responses of the system (it MUST stay within the boundaries of acceptable system states). As Rasmussen, Pejtersen and Goodstein put it:

> "(...) we consider a work system as a functionally coupled entity that adapts to the opportunities and requirements posed by its environment under the control of its human actors (Rasmussen et al., 1994, p. 35)."

The exploitation of opportunities requires a flexible work system. Requirements posed by the environment very often work in the opposite way, leading to higher rigidity of a work system (e.g. regulations, environmental standards, social contracts). Rigid goals and constraints could be formulated as 'functional' in a hierarchical means-end structures. Such 'functional objectives' are implemented in the form of contracts, policies, standards, or technology. Intentions - as the human activity to exploit or use the remaining degrees of freedom - on the other hand are implicit, situational, and require communication[6].

---

[6] Rasmussen, Pejtersen and Goodstein (1994) are not systematical in this point. They use the term 'objectives' both to denominate functional goals and

In the example of the city as a work domain, Rasmussen, Pejtersen and Goodstein separate functionality from intentionality in a means-ends abstraction hierarchy (Table 6).

Table 6: Functionality and intentionality of a city (Rasmussen et al., 1994, p. 43)

	Functionality	**Intentionality**
Purposes and constraints	Goals and objectives	Explore opportunities and constraints
Abstract functions	Implications of the functions in terms of values and resources absorbed	Setting priorities and directing flow of money, people, goods to serve the higher level goals.
General functions	Functions of a city: transport, trade health care, administration, public education	Coordinate lower level processes to serve the various functions
Physical processes and activities	Processes of a city: moving goods and people, sleeping feeding, shaping and assembling products, chemical and physical production processes	Control of the configuration and boundary conditions
Physical form and configuration	Material objects: people and houses, furniture and tools, cars and streetlights	Selection of objects

intentions. They further use the terms 'intention', 'intentional', 'intentionality' (p. 43), 'intentional activity' (p. 48), 'purposive acts' (p. 49) somewhat synonymous. In my opinion, the term 'purposive act' does not make sense, since we should assume that there are no 'unpurposive acts' in a work system. Furthermore, the use of 'intentional' should be limited to the description of activities or structures rather than systems.

"It shows how the functional and material features of the work system dominate the representation at the lower levels, while the intentional features, that is, the objectives that govern the control of the system functions, dominate at the higher levels (Rasmussen et al., 1994, p. 37)."

This is a very interesting approach, but the authors are not further exploring the implications of such a differentiation, except for the classification of work domains[7] on the 'sources of regularity' dimension (Rasmussen 1994@49-54}. Questions regarding the terminology used for the top level arise: Are functional goals and objectives explicit (policies, standards) or implicit (norms, culture)? How do they relate to intentionality on the top level? How do they establish intentionality on this level? Who communicates that and how?

Intriguingly, these two sides of the coin remain somewhat without connection. What is the nature of mediating processes that 'translate' intentions to functions and vice versa? What are the functional aspects that change intentions over time? What or who has agency in their system?

Rasmussen et al. point to the interaction of the two sides only in terms of the importance of the intentional part as a coordinative 'force' on all levels of the material world:

"The intentional component becomes increasingly influential at the higher levels and also more complex as global functions come into play. It can also be seen [in the example of the city] that the intentional part of the domain representation constitutes a hierarchical control function that serves to coordinate the behavior of the material world at all levels (Rasmussen et al., 1994, p. 44)."

According to Rasmussen and colleagues, the analysis of the 'hierarchical control function' - as they also call the intentional part of a work system - becomes more important in systems with sources of regularity other than physical processes (e.g. in manufacturing that involves manual work, companies that provide services, or design) to understand the work domain. But they do not make explicit the

---

[7] "The weight of the intentional constraints compared with the functional, causal constraints can be used to characterize the regularity of different work domains (Rasmussen et al., 1994, p. 49)."

relationship between the hierarchical/intentional control function and operational control (as a mostly technology-driven process).

> "Decision making at each level will deal with discrepancies between the functional state of affairs and the intentions derived from the ultimate goals. Therefore, it is necessary that the representations of the functional implications match the presentation of the intentional explications at each level (Rasmussen et al., 1994, p. 44)."

In the quotation above, the interaction of the control functions is 'locked away' in the black box of human decision making. Here, the missing conceptual definition of 'intentionality' becomes obvious. Two interrelated cognitive activities are mentioned, 'functional implications' and 'intentional explications'. Functional implications could be understood as the result of a cognitive process comparing the actual state of affairs with intended outcomes on different levels of the abstraction hierarchy, thus linking the 'functional domain' to the 'intentional domain'. Intentional explications on the other side are the result of a cognitive process that dynamically updates intentional states (derived from 'ultimate goals') in order to lead or guide activities on within the 'functional domain' - an therefore requiring explicit communication of intentions (e.g. through encouragement, tactical orders, job sequences or schedules). Or, as Rasmussen and colleagues put it - referring to the black box of decision making - intentional explication (as performed through organizational routines and practices) is shaping individual decision making at work:

> "For the individual decision maker, the intentional element is a very real part of the domain. It is embedded as behavior-shaping constraints in the institutional practice and the accepted rules of conduct of the (...) [work] system (Rasmussen et al., 1994, p. 44)."

Interestingly, these intentional elements, and thus the decision-making behaviors, are - according to the authors - being shaped through two features of the intentional structure of a work system:

> "(...) the actual, individual goals of the smaller units are not found by a decomposition of the overall goal but are developed independently from subjective preferences. Basically, this is a consequence of some of the intentional structure of a work system being embedded in the rules of conduct of the social system and

some of it being brought to bear by the individual actors in order to resolve the remaining degrees of freedom (Rasmussen et al., 1994, pp. 45-46)."

Firstly, this has implications for the collection of data: To describe a work domain - including its intentional side - it is necessary to conduct interviews at all levels, to cover both structural aspects, (a) the 'rules of conduct' and (b) the 'individual preferences'. But then, to operationalize and clarify the theoretical concepts, the terminology is used in a somewhat vague way. What is meant by 'rules of conduct'? And, are the 'individual preferences' really due to personality traits (like risk aversion) or rather a local and experience-based perception and interpretation of the state of affairs? - The clarification is important, since neither role identities nor personality traits are within the scope of a work domain analysis in Rasmussen's framework.

Clearly, it seems that these considerations are compatible with critical realist thought, although the differentiation of social structures and individual goals and preferences should be made clearer. However, assuming that both features of the 'intentional structure' can be formulated in a 'de-personalized' way, a Cognitive Engineering approach to support these decision-making activities must consider an 'adequate level' of representation in both aspects of the work domain:

> "(...) the representation of the productive, functional features at each level should be comparable to that of the intentional aspects in order to facilitate decision making and choosing among options (Rasmussen et al., 1994, p. 48)."

Does that imply that the intentional aspects should be represented on all levels in order to make intentions transparent for co-workers and other actors in the system? - Not all of them, since that would be overwhelming for individual actors. But:

> "It is important that the work domain representation captures the basic features of a work system that shape the intentional activity of its staff (Rasmussen et al., 1994, p. 48)."

How are these 'intentional activities' related to activities involving strategic and tactical decisions as well as planning and scheduling activities? - It seems rather obvious that these activities depend on an accurate feedback of system states. But then, there is a twist to that due to the historical perspective of all intentional activities within a work system. What can be expected from the other actors at a specific point in time?

> "(...), regular patterns of response by a work system to work activities are a prerequisite for human actors acting purposefully. Thus actors' opportunities to plan depend on their possessing knowledge about the sources of such regularity or, in other words, the internal constrains shaping the system's behavior (Rasmussen et al., 1994, p. 49)."

Are 'regular patterns of response' the missing link between intentionality and functionality? It could be assumed that this is the case, since control involves both the use and the shaping of these patterns - or the intentional structures that support them.

> "Control of the state of affairs within a work system involves operations on and through its internal constraints and can take place via the causal constraints of the physical part of a system or the intentional structure of the people involved (Rasmussen et al., 1994, p. 49)."

Causal constraints or intentional structures both seem to be 'generative mechanisms' in CR terms. The actors in a work system have to be aware of them, to a certain extent. Intentionality therefore also includes social structures, such as norms set by groups of people, associations and organizations.

For example, there could be a shared belief within the city staff that their city should be as safe as possible for all of its inhabitants. This would shape the intentional structure of the work system in a certain way. Or, a manager could exploit on the historical fact that a 'pattern of response' exists in a company by promising a bonus payment by the end of the year if a certain service level is achieved. In doing so, the intentional structure is shaped to evoke a desired behavior of the work system.

It seems that CWA uses a set of related concepts that are compatible but not necessarily adapted to a critical realist ontology:

- *Intentions* are the conscious results of a cognitive process that involves analytical as well as anticipatory activities.
- *Intentional activities* consist of the evaluation, modification, or communication of intentions.

- *Intentional constraints* are individually known or commonly shared restrictions to action based on intentions.
- *Intentional structure* is the sum of all intentional constraints imposed on a work system at a given moment in time.
- *Intentionality* means all activities within a work system that are related to intentions, including their 'materialization' in technology and other artifacts, their codification in written form (standard operating procedures, instructions), their cultural influences (moral standards, social norms), and their dynamics in time, i.e. their historical development.

Some of these facets of intentionality can be further examined when looking at two different classes of work systems, (a) "mechanized systems governed by instructed rules of conduct (Rasmussen et al., 1994, p. 51)" and (b) "loosely coupled systems governed by actors' intentions (Rasmussen et al., 1994, pp. 51-52)". The following quote shows the interaction of the concepts through an alternative terminology added by myself in box brackets:

> "When such systems are organized according to the 'scientific management' paradigm, operations are centrally planned and the intentionality, that is, the objectives of activities [the intentional structure], are [is] embedded in operating instructions propagating top-down through the levels of the organization. In more modern 'just-in-time' type plants, this central planning with its top-down directive flow is replaced to a large degree by on-the-floor improvisations and adaptations to the immediate situation so that the intentionality [intentional activities] becomes more decentralized. In a way, the high level goals [intentional structure] form[s] an umbrella under which the daily operational decisions can be made (Rasmussen et al., 1994, p. 51)."

Note that according to Rasmussen and colleagues, the intentional structure can be almost fully explicitly formalized in the form of standard procedures and instructions when the nature of the work system allows for it. In such an environment, intentional activities are concentrated in central engineering and production planning departments. Increasingly though, in modern manufacturing environments - due to external demands for flexibility - intentional activities become more and more decentralized, leaving room for local decision making as required in the work process. This in turn is more demanding for local actors:

> "Thus the individual decision makers confronting their part of the work domain will have to shape their activities under the influence of different sources of regularity. First of course, are the laws of nature governing the technical side. Then, depending on the application, the intentional direction will come from some kind of combination of preplanned schedules and formal or informal rules of conduct that shape our own behavior as well as the behavior of the other actors (Rasmussen et al., 1994, p. 51)."

Note that here, in a 'mechanized' context, the authors are not talking about individual influences. They are only referring to informal or formal rules of conduct, introducing the distinction of the two on the fly. Formal rules could be understood to consist of communicated standards or procedures, and even routines. Informal rules of conduct might be non-documented routines or behavioral repertories that have emerged within a 'community of practice' over time (cf. Wenger, 1999).

The second class of work systems[8] - the ones that are more loosely coupled - require more active explication, communication and interpretation of intentional information, i.e. more intentional activities on all levels.

> "Coordination and control of activities [in loosely coupled systems] depend on the communication of company-institutional objectives. The intentionality originating from the interpretation of environmental conditions and constraints by the management (...) propagates dynamically downward and becomes implemented in more detailed policies and practices by members of the staff. Making intentions operational and explicit during this process requires an interpretation considering a multitude of details dictated by the local context (Rasmussen et al., 1994, pp. 51-52)."

This imposes a very challenging cognitive demand on the members of an organization situated somewhere in the middle of the hierarchy (i.e. the so-called 'middle management').

> "Therefore, many degrees of freedom remain to be resolved (...) at intermediate levels of an organization. This in turn implies that the individual actor faces a work environment in which the regularity

---

[8] e.g. Hospitals, service businesses, manufacturing plants based on manual work and/or 'just-in-time' production concepts

> to a considerable degree depends on the intentionality brought to bear by colleagues (Rasmussen et al., 1994, p. 52)."

Considering the (known/believed) intentions of the other actors in the system adds considerable complexity to the problem solving and decision making tasks of these managers[9]. But these members of a work organization also have various possibilities to change the intentional structure throughout the organization.

> "To an actor, coping with a complex system, the properties to consider for controlling its state of affairs will be a varying combination of functional and intentional relations. Sometimes, actions will be aimed directly at its functional state. However, control often requires an influencing of other actors' intentional states or, in technical systems, entails modifications of the intentional structure embedded in the control system (Rasmussen et al., 1994, p. 52)."

In an environment with predominantly direct face-to-face interaction, intentions can be made explicit and aligned through a process of negotiation or mediation as conflicting intentions become evident. Such reproductions and transformations of social structures fit well with Bhaskar's model of social activity (see above). But, in a highly distributed work environment involving computers as collaborative tools, these dynamics within the 'intentional domain' are hindered substantially, according to Rasmussen and colleagues.

> "Therefore, (...), when the interaction with a work system is mediated through an information system, constraints originating in the intentional structures, as well as those based on functional, causal relations, must be represented (Rasmussen et al., 1994, p. 53)."

But how can such constraints be represented? How can we formalize intentions on various levels of organizations? And, who needs to know about them? Accordingly, at the end of their book, Rasmussen and colleagues ad an outlook to future considerations to their conceptual elaborations above.

> "We have not been able to find 'ecological' displays for analysis and evaluation of work system states for autonomous system users in a

---

[9] In a study involving expert poker players, Bina, Chen & Milgram (2008) have been able to show the influence of this kind of considerations on the decision making process.

constrained environment (...), that is, for work domains in which the source of regularity is related to intentional structures, such as laws, regulations, and company strategies. The reason for this situation is that *symbolic representations of intentional structures have not as yet been developed*. Activities, such as production flow control in just-in-time production systems, patient treatment planning in hospitals, or case handling in public service institutions, are handled by autonomous actors within the constraints posed by regulations, by the intended product-case flow, and by the intentions of colleagues handling the state of affairs up- and down-stream of the flow of work items. For the design of ecological interfaces, we need to formalize a description of the different forms of intentional structures and to find suitable symbolic representations. This is an area for further development (Rasmussen et al., 1994, pp. 330, italics added)."

What then would be the requirements for the formulation of a more formal description of intentional work domains (without trying to come up with a means-ends-hierarchy)? How can we talk about social 'mechanisms' and structures that form and constrain the work domain? Is it possible to come up with a similarly formal way of describing social realities that are restricting individual agency in a comparable way to causal structures? Is the reality of functional structures (as described in abstraction hierarchies) equally or more justifiable than the reality of social/intentional structures (as described with other methods that need to be developed)?

It's quite obvious that a critical realist ontology will help us to clarify some of these issues (cf. Reed, 1997).

4.4.3 Implications for a critical approach in Human Factors

What are the implications from a critical realist perspective for a critical approach in Human Factors? - Considering the elaborations above, I see the following potential for modification and adaptation of the CWA framework:

- The classification or typology of work domains should no longer be used. It was developed historically but has a limited value in todays work domains.
- The dichotomy of intentionality vs. functionality is ontologically confusing and should be replaced.

- The term 'intentional structure' should be replaced by 'social structure' (position-practice) and 'individual goals' or intentions (i.e. cognitions based on experience).
- Individual behavior is shaped by social structures in every work domain, with no exceptions. Professional associations, unions, teams, units, departments, etc. are all forms of social structures with specific powers and tendencies to influence individual and group activities.
- Work domain analysis needs to be extended, since social structures *and* technology are a substantial part of the work domain. The difficult part hereby is how to analyze and describe the tendencies that are inscribed / materialized in technology based on position-practices or norms of the social groups that have contributed to its development.
- The analysis of events becomes more important: What kind of generative mechanisms could have contributed to them? - A framework and formal language is needed to describe and possibly visualize contributing tendencies.
- Three domains need to be covered: The individual (e.g. strategy analysis), the social and the technological-material. They are more and more overlapping, there is no 'hierarchy' or structure (cf. Vicente, 1999, p. 115). All of them - as a totality - make up the 'ecological environment' of work.
- Alternative causes that could have contributed to the event, or to the prevention of it, should be discussed. There could be tendencies that 'neutralize' each other. Or, there could be a tendency that 'silences' other, weaker tendencies.
- The development of a self-critical element within CWA, an approach to open up for interdisciplinary debate *about* the approach and the methods that are applied.

What are possible methodological candidates to achieve these modifications and extensions, i.e. primarily the formal description and explanation of tendencies (intentional/social structures), exist in literature? The next chapter introduces different approaches that have the potential to describe and formalize such structures in different work environments and finally leads to a selection of approaches and methods for the subsequent application in two case studies.

4.5 Development of an analytical approach

A very basic requirement for the analytical approach I have in mind is its correspondence to realist ontology. It must be able to shed light on causal powers, tendencies and liabilities, and the multiple determination of events. Furthermore, it should be supporting retroduction and retrodiction in modeling and theory building: "... real causal powers identified by retroduction become building blocks in the retrodictive construction of explanations of actual events (Elder-Vass, 2010, p. 48)". Finally, there must be a basic openness for the identification of complementary contributions: "Bhaskar's model of actual causation and multiple determination provides a framework for constructing causal explanations that recognize the complementary contributions of emergent properties at a variety of different levels (Elder-Vass, 2010, p. 53)." Thus a laminated view of different levels of a 'thing' or organism should as well be supported. Any kind of 'conflation' or overtly restrictive 'closure' must be avoided.

In sum, what is needed for our sought after theory/methods package is:

1. A critical realist *local / scientific ontology* for Human Factors as a template or background, covering both technological and social spheres,
2. a *set of adapted / extended analytical methods* to develop local theories with the help of retroduction and retrodiction and
3. a *self-critical, incremental research process* that allows for validation and evaluation of the explanations and theories that we develop.

One of the many questions that rise from these required features is: Which empirical methods could help us to 'retroduct' and 'retrodict' explanations within work domains of interest? - Such methods should be sufficiently powerful to address the problems I have identified in chapter 3. Several methods have been developed in the last three decades that take a more or less critical approach to the analysis of social and technical phenomenon. Some of these, such as Ethnomethodology, Actor-Network-Theory or post-structuralist theories have been widely criticized in organization or technology-oriented sciences in recent years, mostly because of their 'flat' ontology and confluence of structure and agency (cf. Elbanna, 2006; Mutch, 2002; Rammert, 2007; Reed, 1997; Rose & Jones, 2005;

Vandenberghe, 2002). As a consequence, I will *not* consider these methods for my approach.

However, there are two methods that could be feasible indeed, for that they afford more 'analytical dimensions' and allow for a distinction of agency and structure. One is the analysis of discursive practices through *socio-cognitive discourse analysis* (Elder-Vass, 2012b; Keller, 2007; Lacau & Mouffe, 2001; Parker, 2005a; van Dijk, 2008a; van Leeuwen, 2008a). The second is Archer's *morphogenetic approach* (Archer, 1995, 2002; Elder-Vass, 2007; Mutch, 2010). In order to prepare the application of these two methods on the empirical material of two case studies, I first want to investigate the possibilities of a specific *regional ontology* for Human Factors, since the ontology will provide the most important theoretical fundament for the discussion of the utility of these analytical tools that comes later on.

4.5.1 Development of a regional ontology for Human Factors

Based on Elder-Vass' method for the development of regional ontologies and his social ontology (Elder-Vass, 2010, 2012a), I intend to develop a regional/scientific ontology for Human Factors. This ontology must include not only technical-material structures but also social structures. Rasmussen, Pejtersen and Goodstein (Rasmussen et al., 1994, p. 26) might have had something similar in mind since they argue for a dual approach, hereby combining the 'identification of activities' and the 'identification of actors'. Both are essentially intended to serve the same goal, that is a description of the functionings of a work system that is not normative but an approximation to the 'real'. Especially the proposed analytical categories of 'role allocation' (social structure) and 'management style' (an emerging property of a social structure) will most likely be part of that envisioned ontology.

To develop a local ontology - according to Elder-Vass - we must identify the following aspects by mapping the theoretical knowledge of the discipline onto a realist ontology:
- the particular types of *entities* that constitute the objects of the discipline,
- the *parts* of each type of entity and the sets of *relations* between them that are required to constitute them into this type of entity,
- the *emergent properties* or *causal powers* of each type of entity,

- the *mechanisms* through which their parts and the characteristic relations between them produce the emergent properties of the wholes,
- the *morphogenetic causes* that bring each type of entity into existence,
- the *morphostatic causes* that sustain their existence,
- and the ways that these sorts of entities, with these properties, *interact to cause the events* we seek to explain (Elder-Vass, 2010, p. 69).

Regional/scientific theory is concerned with identifying the causal mechanisms underlying emergent properties (retroduction), and with explaining how these interact to produce events of interest (retrodiction) (Elder-Vass, 2010, pp. 72, citing Lawson). Retroduction requires the use of established methods for empirical research, including both quantitative and qualitative techniques.

The general framework approach of CWA (Vicente, 1999, p. 136) fits well into a critical realist research strategy. First of all, we need 'conceptual distinctions' (a local ontology) to realize 'modeling tools' (retroduction and retrodiction) that help us to develop 'models of work constraints' (local theories about emerging properties and causal mechanisms) which inform design interventions. CWA's hierarchical abstraction and decomposition approach - known as 'work domain analysis' is a good starting point: The principle of decomposition of the work domain into its entities is certainly compatible with a realist ontology. The question of emergent properties needs to be resolved though. Instead of using the function (means-ends) abstraction hierarchy (Rasmussen et al., 1994, p. 36ff; Vicente, 1999, p. 149), a work domain analysis could be envisioned as a kind of local ontology. The kind of means-ends abstraction hierarchy that works for technological systems might not be adequate for all kinds of work domains (Vicente, 1999, p. 164).

The analytical decomposition of work domains that was originally proposed by Rasmussen and colleagues (Rasmussen et al., 1994, p. 36) was intended to explain problem-solving behavior by experienced technicians. Therefore it combines reasons and ultimate goals with the physical components of a system at the other end. To talk about *reasons* is not compatible with a realist ontology though. There are emergent properties of entities and there is individual behavior with intentions, plans or goals. Reasons are things that reside within individuals, and there is a discussion going on whether or not reasons can have causal powers (Elder-Vass, 2010, p. 93ff).

Critical realist work domain analysis would therefore consist of the formulation of a local theory based on a realist, laminated *sociotechnical ontology*. When considering Rasmussen, Pejtersen and Goodstein's 'classes' of representations within a work domain (Rasmussen et al., 1994, p. 38) it becomes clear what the differences would be. Their hierarchy is a fusion between a physical-logical decomposition and a goal hierarchy. This works well for predominantly technological systems, since distinct parts of a system were designed to fulfill a certain purpose, e.g. the cooling system of an engine. When it comes to more complex work domains, there might be various entities involved in producing an effect, such as a sales team and a Customer Relations Management system that achieves a certain quality of service to the customer.

Therefore I propose to use the social ontology developed by Elder-Vass (2010) for work domain analysis. It focuses on groups as *normative circles*. Groups or normative circles are central to the social structures that make up organizations and society as a whole. In this social ontology, *normative social institutions* are emergent properties of normative circles (Elder-Vass, 2010, p. 122). Each member of a norm circle holds a normative belief or disposition endorsing a certain *practice*. They need to be aware - at some level - that they are expected to observe the norm, and that they face consequences when they do (positive) or do not (negative) comply with it. Normative institutions contribute to individual behavior, through the commitment to endorse and enforce a practice by the members of a norm circle. They share a collective intention to support the norm, although the do not necessarily share the norm in private. As Elder-Vass puts it: "This (...) leaves open two important possibilities: (a) that conformance with norms may sometimes be a consequence of prudential behavior in the face of unequal power relations rather than consensus over the value of the norm; and (b) that members of the norm group who disagree with its standards (even if they do actually conform with them) may take action directed towards changing those standards (Elder-Vass, 2010, p. 126)."

Organizations - according to Elder-Vass - are structured social groups with emergent causal powers. They depend on normative mechanisms to produce the *role specialization* upon which they depend, but *role-coordinated interaction* between their members (which may include non-human material things) provides a further class of mechanisms, a class that confers non-normative causal powers on the organizations concerned (Elder-Vass, 2010, p. 167). Authority relations are a variety of role specialization, they confer some part of the power of the organization as a whole on certain persons (as role occupants). The management role includes the development of role specifications in response to the goals, performance and circumstances of the organization.

In addition, I propose to use an ontology of technological systems that takes into account their real powers as social structures, for example though the distribution of user roles and access rights to data or content. Technological systems are physical as well as 'institutional' in the sense that they are social structures of a material kind. They embody norms of behavior, degrees of freedom or constraints to agency that have been inscribed to them by their designers, who in turn have been influenced by their own social environment.

However, a simple "return to the material (Kallinikos, Leonardi & Nardi, 2012, p. 3)" will not be sufficient. As Kallinikos and colleagues are discussing, it is impossible to distinguish between the material and the non-material. For example, the mathematical and algorithms concepts behind production scheduling are not only inscribed into a software program (material) but serve also as psychological tools for the planner, who is using these concepts in his everyday work. The more the mental structures of the persons involved are adapted and aligned to the material and corporate world around them, the more entangled both worlds get, the more difficult it becomes to analyze and describe their individual nature.

4.5.2 Socio-cognitive discourse analysis

Next I will present and discuss socio-cognitive discourse analysis as a possible analytical tool that fits to a critical realist perspective. There is a variety of approaches and methods to discourse analysis available in literature to empirically study these 'modes of rationality' or discourses (cf. Keller, 2011). I chose critical or socio-cognitive discourse analysis (van Dijk, 2001; 2002; 2008a; van Leeuwen, 2008a; Wodak, 2006) because of its explicit use of critical thought, its emancipatory objectives, its empirical fundament and its self-reflecting approach to scientific practice. Also, there seems to be a fit to our ontological template (cf. also Elder-Vass, 2012b): "Rather than merely describe discourse structures, it tries to *explain* them in terms of properties of social interaction and especially social structure (van Dijk, 2001, pp. 353, italics added)."

As Clegg and Wilson (1991) have observed, there is a plurality of ongoing discourses or - in their words - *'modes of rationality'* within organizations. In their perspective, all agents in and around the organization are developing diverse and simultaneous rationalities that are in a constant flux. These modes of rationality "are built out of locally available conceptions which embed economic action (Clegg & Wilson, 1991, p. 247)." This particular knowledge may be derived from local custom or practice, both having been shaped by organizational culture and institutional framing. Not all agents share the same knowledge and rationality.

This view relates to the notion of local *group ideologies* that not only reflect general streams of ideological discourse, but also are distinctive for a certain subset of organizational members with common interests. As van Dijk puts it: "(...), we need to attend primarily to those properties of discourse that express or signal the opinions, perspective, position, interests or other properties of groups. This is specifically the case when there is a conflict of interest, that is, when events may be seen, interpreted or evaluated in different, possibly opposed ways (van Dijk, 1995, p. 22)". Such representations of ideologies are often articulated along an *us* versus *them* dimension, and therefore any property of discourse that expresses, establishes, confirms or emphasizes a self-interested group opinion, perspective or position is deserving attention in our analysis.

The elements of local rationalities as they are revealed through linguistic analysis are indicators of actual ways people are mentally representing their environments. This process could be described as the construction of mental models of the situation - also called *context models* - based on previous experience but also on influences of ongoing *ideological control*. It is through living discourses that are predominant in a certain social environment that such cognitions are formed. And on the other hand, these discourses are reproduced and modified through such social cognitions (van Dijk, 1995; 2008a; 2008b).

According to van Dijk, discourse and context are interrelated. Contexts control the production of discourse and comprehension (van Dijk, 2008a, p. 16). In his theory of context, contexts are mental models representing communicative situations. As such, they are a special type of mental models of situations and everyday experiences, which he calls 'experience models'. Although contexts are individual and unique, as subjective definitions of communicative situations, they have a basis of shared social cognitions. Contexts are both personal and social at the same time. The fundamental function of context models is to make sure that participants are able to produce text or talk *appropriate* to the current communicative situation and understand the appropriateness of the text or talk of others (van Dijk, 2008a, p. 18). The methodological implications are as following: "... we try to explain, within a theoretical framework, why specific discourse structures are being used and not others. Thus, by some kind of psychological or methodological reverse-engineering, we may go back from discourse properties to probable context-model structures, event models, and their underlying belief systems, each related to the situational and social structures as known and perceived by the participants (van Dijk, 2008a, p. 224)". Discourse is understood as a specific communicative event, an oral or written form of verbal interaction, or non-verbal expressions such as pictures and gestures (van Dijk, 2008b, p. 104). These interactions can be described through structural analysis.

In his approach to discourse analysis, van Leeuwen uses the discursive re-contextualization of social practices as a starting point (van Leeuwen, 2008a). Social practices are socially regulated ways of doing things. The regulation can be achieved in various degrees and in different ways, for example through strict prescription, traditions, the influence of experts or charismatic role models. Social practices include a set of elements, which are not always represented in total: Participants, actions, performance modes, eligibility conditions, presentation styles, times, locations and resources. What is providing the leverage point for analysis is the transformations that such social practices go through when re-contextualized in speech or written form. Van Leeuwen proposes substitutions (e.g. participants are particularized or generalized), deletions (e.g. actions or resources), rearrangements (e.g. reverse order of actions) and additions (e.g. repetitions, reactions, legitimations, evaluations).

If we consider our field of research, such theoretical framing seems compatible with our objectives, that is, to 'open the black box' of technology and analyze the duality of technological and social structures in depth. In order to do their job, planners and schedulers need to mentally represent or model the entire manufacturing environment. This includes physical features like machine capacities and bottle-necks, but also social factors like motives or incentive structures. Combined, physical and social elements form *social practices*. Production planning and scheduling itself *is* such a social practice; it is a *socially regulated* way of doing the work of *planning*. ERP technology is enabling, constraining or restricting this particular social practice. Through deployment and implementation of off-the-shelf technology social practices are formalized and incorporated into the system in more or less consistent ways. This process of configuration is subject to discursive *re-contextualization* of social practices, which are effected through more or less transformed representations (van Leeuwen, 2008a). Increased dynamics of such transformations have been observed within ERP implementation projects (Hanseth et al., 2001). An analysis of discursively realized recontextualizations therefore offers further insight into the characteristics of context models that constitute local rationalities.

In sum, the socio-cognitive approach to critical discourse analysis offers an analytical approach and some methods that are coherent with a realist ontology. It provides a theoretical framework to connect social interactions with social structures through a detailed analysis of communicative events. As such it will help us to fill some of the blank spots in our explanatory theories about the implementation and use of ERP systems. To achieve this, the interpretation and analysis of empirical material such as observation protocols, interviews and visual material such as advertisements will be necessary (cf. Keller, 2011, p. 276).

In a very much related approach, Jian (2011) applied a discursive framework of organizational changing in an interesting analysis of discourses and 'local rationalities' in an insurance company. In his framework (cf. Figure 9}, organizational change is discursively constituted through a variety of articulations concerning organizational circumstances and identities. Interestingly, this framework allows for the conceptualization of the influence of

A critical theory/methods package for Human Factors

normative circles such as 'managers of a certain school' or 'most-competitive organizations'.

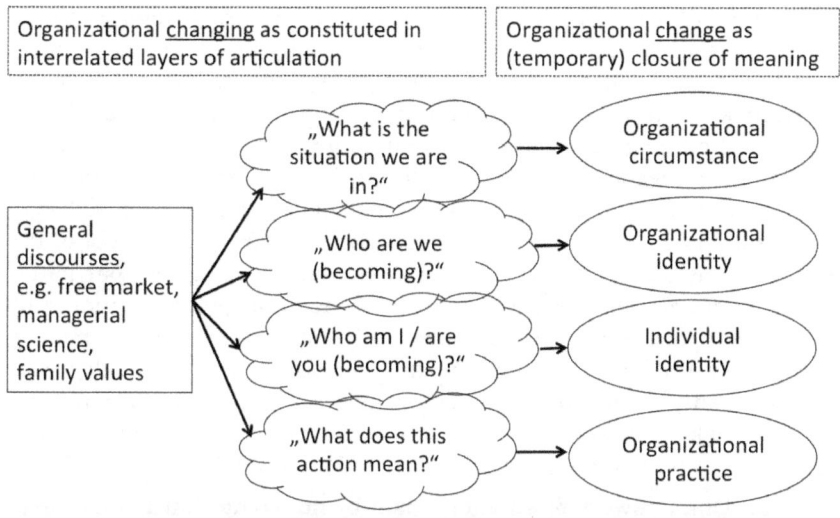

Figure 9: A discursive framework for organizational change (adapted from Jian, 2011, p. 47)

According to Jian, one or more layers of articulation can be identified in a specific communicative act. He also stresses the preliminary nature of any fixation of meaning that is achieved through the 'struggle' of discursive articulations within an organization. This framework underpins the dynamic nature of organizational discourse and by doing so points to a weakness of socio-cognitive approaches to discourse analysis: Their failure to deal with intersectionality (for a critical realist account of intersectionality see Elder-Vass, 2010, p. 131). I will pick up this interesting approach later in chapter 5.4 to analyze an advertisement by a vendor of an ERP system.

A very similar critique and extension or addition to critical realist thought is made by Willmott (2005): Critical realist perspectives on control in organizations should avoid to simplify relations of subordination within organizations. It is not sufficient to put capitalist interests as the only 'causal power' that defines work organizations or economic spheres. Interests should be seen as socially organized and identified rather than imposed as an external force.

Jones (2007) criticizes discourse analysis that aspires to be critical but embraces a predominantly linguistic approach without further political engagement. In his view, substantial critique is impossible when sticking to linguistic categories that are incapable of grasping the fine grained practical and political implications of what is said. What is needed according to Jones is engagement in moral and political terms: "As with any form of behavior, our critical appraisals of and reactions to communicative practices spring from our sense of the personal, institutional, or political rights and wrongs of particular engagements and our feelings about how such engagements should be conducted. That is why we can also, at least in principle, be held to account for the treatment we are dishing out when we speak up or remain silent, when we interpret or mis-interpret what others say, when we ask questions and give answers, when we make requests or give orders, and when we comply or defy (Jones, 2007, p. 343)."

Van Dijk answers to such criticism by stressing that his aim is to combine both, sociological (macro) analysis and linguistic (micro) analysis: "It is precisely in these macro-micro links that we encounter the crux for a critical discourse analysis. Merely observing and analyzing social inequality at high levels of abstraction is an exercise for the social sciences - and a mere study of discourse grammar, semantics, speech acts or conversational moves, the general task of linguists, and discourse and conversational analysts. Social and political discourse analysis is specifically geared towards the detailed explanation of the relationship between the two along the lines sketched above (van Dijk, 2002, p. 83)."

When positioning socio-cognitive discourse analysis, therefore, we would chose a place somewhere in the middle between macro- and micro-levels of sociology. This fits well with our local ontology since it potentially supports the detection and description of role specifications and role-coordinated interactions within organizations. An important aspect that is missing form this kind of analytical framework is that of material 'things', since it is predominantly used to analyze all kinds of *text*. In addition to that, "... the existing critical social science work on information systems both downplays the influence of technology supply and often overlooks the influence of the broader historical setting on the unfolding of the [ERP] technology (Pollock & Williams, 2009, p. 276)". Therefore I would like to consider a second analytical approach, based on the theoretical body developed by Archer (1995).

4.5.3 Morphogenetic approach to organization and technology

The morphogenetic approach to realist social theory building was initially developed by Archer (Archer, 1995, 2002). It builds upon Bhaskar's critique of the social (and psychological) sciences (Archer, 1998; Bhaskar, 1979). Archer's approach could be called emergentist, in contrast to what critical realists call methodological individualism (Elder-Vass, 2007). The focus on morphogenesis and emergence in the interplay between (human) agency and (social) structure allows for a distinction of the two domains – in opposition to or distinction from Gidden's structuration theory and its derivates (cf. Mutch, 2013). Emergence is the key to the fundamental ontological claim of social realism: That social structures, although the product of human individuals, have causal powers of their own, which cannot be reduced to the powers of those individuals (Elder-Vass, 2007; 2010; 2012a).

Besides the use of emergence as a principle for explanatory theories in critical realism, there is another principle, historicity. According to Archer (1995), confluence of agency and structure can only be avoided through a historical lens. This encompasses the notion of three temporally distinct phases that are analytically relevant: (1) Structural conditioning that is essentially constraining performance, (2) social interaction and actualization of causal mechanisms at the moment of enactment and (3) structural elaboration or modification and thus the reproduction of social structures through these activities (Figure 10).

In order to comply with critical realist or morphogenetic theory, "... a complete causal analysis of the real powers or emergent properties of any emergent entity would include five elements: (a) a list of its characteristic parts; (b) an explanation of how these must be structured (i.e. related to each other) to form the whole; (c) a morphogenetic account of how this comes about; (d) a morphostatic account of how it is sustained; and (e) an explanatory reduction showing how the powers or properties of the whole are produced as a result of it having the parts it does, organized as they are – in other words, an explanation of the generative mechanisms underlying each causal power (Elder-Vass, 2007, p. 31)."

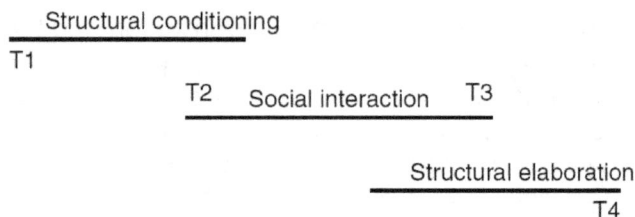

Figure 10: The morphogenesis of structure (adapted from Archer, 1995, p. 193)

Vokoff and colleagues (2007) found that critical realism provides an appropriate lens for examining the actions and interactions of stakeholders in their appropriation and use of technology. It also incorporates a temporal aspect, which fits well with change processes related to the implementation of large-scale information technologies in organizations. They argue that critical realist thought fits well to a grounded theory approach to empirical study, since both lead to explanatory process theories that explain events not by predicting what will happen nor describing what did happen. Such theories are focussing on identifying the mechanisms that generate what is observed in the empirical domain. Doing so, they avoid neglecting the role of technology: "By using a critical realist lens instead [of a structurationalist lens], we are better able to address the inherent materiality of technology (Volkoff et al., 2007, p. 833)".

To understand the importance of an ontology-driven approach to technology in organizations it is helpful to consider organizational routines as one of the main characteristics of organizations. During the past decades, a vast amount of scholarly work has been conducted on this topic (for an overview, cf. Becker, 2004). Ontologically, Feldman and Pentland (2003) propose a distinction between an ostensive (or structural) aspect of routines and a performative (or agential) aspect. Although their proposition is strongly informed by

Gidden's structuration theory, they insist that the specific nature of organizational routines justifies a specialized terminology. Besides the double nature of organizational routines, they define them as repetitive, recognizable patterns of action with multiple participants and interdependent actions (Feldman & Pentland, 2003, p. 104; Pentland & Feldman, 2005, p. 795). They understand organizational routines as 'generative systems' (cf. Figure 11).

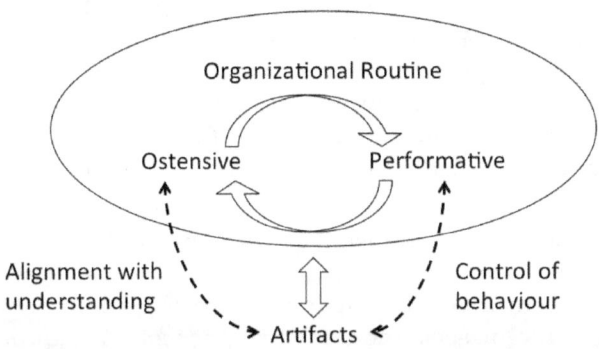

Figure 11: Organizational routines are generative systems (adapted from Pentland & Feldman, 2005, p. 795)

Using the example of organizational routines, it becomes evident that these collective enactments are emergent properties or powers of specific ensembles of persons (role incumbents) and material (buildings, computer systems, tools etc.). In order to be performed, organizational routines depend on social interactions of humans with specific knowledge and roles as well as structurally elaborated environments. The performance itself is inherently improvisational. There is a certain flexibility that allows for the transformation of the routine (its ostensive aspect), but within limits (mostly due to its material aspect, cf. Volkoff et al., 2007, p. 845). This could be termed potential for endogenous change (Feldman & Pentland, 2003, p. 113). Whilst being flexible in principle, organizational routines are heavily involved in the reproduction of social structures within the organization. Members of the organization are generally not allowed to deviate substantially from the routines, there might even be sanctions in place to make sure that employees are compliant with the prescribed pattern of actions (e.g. standard operating procedures). In terms of the morphogenetic approach, organizational routines are contributing to the morphostasis of an organization.

Extending this view on organizational routines and information technology, Mutch (2010) proposes to focus on the architecture of ensembles as they are defining the scope of a technology. Specifically, he argues that students of organizations and technology have to put more emphasis on the features of technology, especially the emergence of data structures from particular combinations of hardware and software. According to Mutch, these structures are relatively enduring and are shaping the position-practices of agents within the organization. As opposed to groupware or other more customizable software products, datawarehousing systems – Mutch's example – are requiring substantial expertise and financial resources to establish and operate.

The notion of technology as inscribed structure is looking at a long history, especially in the tradition of socio-technical systems (Mutch, 2013). In recent times, inscription processes have received a lot of attention, mainly in social constructivist schools, i.e. through the social construction of technology approach. However, not enough attention has been allocated towards the relationships between inscribed structure and the performance of organizational routines as 'sociocultural action', followed by elaborations and modifications of the structure (Mutch, 2010, p. 511). In order to achieve a more comprehensive understanding of technology, it is necessary to look closer at the ostensive aspect of what people think about an organizational routines, and the ways the 'best practice' is implemented in a technological structure. Pentland and Feldman are distinguishing between two analytical aspects (cf. Figure 11): "While the relationship between artifact and performance is about the control of behavior, the relationship between artifact and ostensive aspect is about an alignment of documents and other objects with what we understand about what we are doing (2005, p. 807)." What is crucial here is that there might be different views on how to do things within the company. The 'best practice' – possibly already inscribed in the technology (Volkoff et al., 2007) – might be in line with the view of some managers, but not others. As Pentland and Feldman put it, "...it is not uncommon for there to be no consensual understanding or for the consensual understanding to be different from the artifact (2005, pp. 807, italics added)". Differences in understanding of work practices might also reflect cultural backgrounds or shared social norms within certain groups.

4.6 Discussion of proposed theory/methods package

It is my intention, as described above, to propose a kind of theory/methods package for a critical Human Factors approach. How far have I come in this endeavor?

On the theoretical side, I have identified Critical Realism as a potential source of orientation and inspiration. The formulation of a sound theoretical fundament in the form of a local/scientific ontology for Human Factors provides a starting point for further epistemological and methodological considerations, elaborations and discussions.

In the following, I have proposed to reformulate some of the principles of Cognitive Work Analysis as an Human Factors framework in order to open up for additions to the CWA toolset. In particular, work domain analysis needed some attention in its capacity to handle 'intentionality' and 'social structures'. It seems that both the framework and the toolset could be updated or extended to be ready for a test on empirical material.

Further, I have identified two methodologies that seem to provide a compatibility with critical realist thought as well as with employability for a mixed social and technical research topic such as ERP systems. The more limited but also more feasible approach is socio-cognitive discourse analysis, the more demanding but also more powerful explanatory route is the morphogenetic approach. Both allow for a self-reflexive, iterative research process that acknowledges the preliminary nature of explanatory process theories.

What follows in chapters 5 and 6 are two case studies on computer-supported planning and scheduling in manufacturing as illustrative examples on how the selected theory/methods package(s) can be put to work. Both chapters consist of a description of the study approach and design, followed by an analysis of the empirical material using the two selected methodologies described above.

5 Study one

> Though scarcely recognized, sense and reference are central questions of contemporary work and life. Their significance in future systems of production, consumption and administration will certainly rise as the world is increasingly transformed into a vast electronic landscape (Kallinikos, 1995, p. 139).

The case study has been conducted in a Swiss manufacturing company using a set of qualitative methods to investigate and model the structure and behavior of the planning and scheduling work system and discuss issues of the actual socio-technical system in place. The approach was derived from an established human-technology-organization perspective, called M-T-O analysis, that had been developed at the Swiss Federal Institute of Technology in Zürich in the 1990s (Grote et al., 1999; Strohm & Ulich, 1997; Wäfler, Windischer, Ryser, Weik & Grote, 1999). The general intention of this methodology is to analyze an organization as a whole, from organizational to technical and to psychological aspects of the work environment. At the core, it serves the establishment of meaningful, comprehensive work mandates[10] for individuals and groups - coming form a tradition of humanist work psychology and a substantial body of research providing evidence for organizational and psychological benefits stemming from this way of organizing (cf. Hacker, 1995; Trist & Bamforth, 1951; Ulich, 2005). As a methodological framework, M-T-O analysis makes use of several established psychological research instruments, such as VERA (Oesterreich, 1991), KABA (Dunckel et al., 1993), SALSA (Rimann & Udris, 1997) or KOMPASS (Grote et al., 1999).

M-T-O analysis consists of eight steps. The first step is an analysis of the products and the general production environment of the company. The second step involves the following-through of prototypical orders, to assess functional integration, planning quality,

---

[10] As a translation of 'Arbeitsaufgabe', I prefer to use 'work mandate' instead of the more general, but also unspecific 'task'. The reason for this is that a task, like in task analysis, is generally more related to single work steps ('Handlungen' or 'Operationen', cf. Hacker, 1995).

number of interfaces, quality of interfaces and unnecessary redundancies. In a third step, all departments are rated according to their independence, their coherence of mandates, the correspondence of product and organization, the polyvalence of employees and the technological-organizational convergence. The fourth step involves the analysis of groups (work mandate, temporal resources, work conditions, qualifications, quality, internal and external coordination, decision latitude) and the fifth is focussing on key tasks. The sixth is targeting the individual worker and his or her subjective assessment of the work situation. The last two steps are devoted to a historical analysis of the sociotechnical design. After a general historical overview, one or two milestones of the socio-technical development are analyzed, for example the introduction of a PPS or CAD system. This socio-technical analysis can be complemented with a individual/psychological analysis of work mandates (Dunckel et al., 1992; Hacker, 1995; Matern, 1984; Volpert, 1987).

5.1 Access and data collection

The company that provided the environment for this case study is a mid-size Swiss manufacturing company, employing roughly 8000 persons, mainly in Switzerland. The investigation took place at a time when the company was preparing to implement a new enterprise resource planning system. Just before the year 2000, a previous system had been installed under great time pressure - for there was a fear that the old computer system would not work after 'Y2K'. However, the company was never fully satisfied with that ERP system. Therefore the management had decided to replace it with a more widely used and highly customizable new system. A project team was at the time of the study establishing the requirements for the new system, and proposing possible solutions to existing problems. The project manager was an information systems engineer from the IT department. We as researchers were not directly involved in the project but were invited to do our study and finally present the results of our analysis to a selection of interested parties and stakeholders within the company.

The company had introduced an enterprise resource planning system in 1999. At the time of the field work it was in the process of replacing this system with a new one from a different vendor.

Discussions about features of the new system and the exact modalities of its implementation were under way, and therefore the company provided an excellent environment for the study of pre-implementation discourse concerning enterprise resource planning systems. We contacted the company because we were interested in a case study of ERP implementation. The initial contact was established through the quality manager as well as the chief financial officer of the company. Our affiliation with the University of Applied Sciences Northwestern Switzerland allowed us to enter the company without conflicting interests. The only implicit motive on our side was an interest in a sustainable relationship with the company to provide more opportunities for research. The senior management agreed to our research approach because it was convinced that our results would complement to the ongoing IT requirements engineering process. At the time of our study, the central IT department was analyzing requirements using its own resources.

The company's production of high-precision parts for the watch industry involves a large amount of different products and a substantial variety within products (variations). The production planning involves up to nine levels of parts, sub-assembly and assembly groups. The planning process is distributed across a central product-oriented planning department, a local planning and logistics department in the factory, and among production unit leaders on the shop floor. Planners and schedulers from all levels were involved in the study, plus members of management and the IT department. The methods we used were based on an extension of the human-technology-organizations framework, as described in an internal report by Gasser, Gärtner and Wäfler (2008) as well as in a recent book chapter by Wäfler, von der Weth, Karltun, Starker, Gärtner, Gasser and Bruch (2011).

After an initial analysis - together with the central IT department - of the situation and the production planning process, the following levels of the organization were identified for a more detailed analysis: Centralized planning, decentralized planning and scheduling, and shop floor dispatching/supervision. An additional interview with the factory manager of the chosen production unit was scheduled to better understand the general situation of the factory within the overall company. The field work for this study was therefore mostly done in one particular manufacturing unit. The focus hereby was on operational uncertainties and control capabilities to cope with them on different levels of the planning hierarchy.

A series of ten semi-structured interviews was then conducted over a period of six weeks with seven interviewees in total. The interviews covered standard topics around socio-technical issues of production planning. They typically lasted between one and two hours and were recorded and transcribed (cf. Table 7). A tailored guideline served as a reference for the interviewer. For that purpose, a set of guidelines was developed following the principles of M-T-O analysis (Gasser et al., 2008). Table 8 is showing one of these guidelines, used in observation interviews with persons that are predominantly doing planning type of work.

Table 7: Interviews and observations in case study one

Nr.	Department / role	Duration of Interview	Topic	Duration of observation
1	Information Technology	1:51	Work systems	-
2	Centralized Planning	2:10	Operational uncertainties	-
3	Centralized Planning	0:30	Operational uncertainties	-
4	Factory management	0:53	Work systems	-
5	Shop floor supervisor	0:49	Work systems	-
6	Decentralized planning	2:42	Operational uncertainties and control tasks	-
7	Centralized Planning	0:42	Operational decisions	3:15
8	Decentralized planning	1:02	Operational decisions	3:45
9	Centralized Planning	1:58	Operational decisions	3:45
10	Centralized Planning	1:52	Operational decisions	3:45

Table 8: Observation interview guideline for persons with planning tasks (Gasser et al., 2008)

Questions to ask persons with planning tasks
When is a new production order issued (event / trigger)?
Which parameters are set in doing so (e.g. amount, deadlines, resources, operations)?
Identify for each parameter: Uncertainties, support, interferences etc.; possible questions: - Which objectives are considered when setting the parameter?   - Which decisions do you have to take?   - Which is your decision latitude? What are the alternatives? How (with which rules) are you choosing between the alternatives?   - What is relevant information for the decision?   - Where (from which source) does that information come from (e.g. ERP system, meetings, discussions)?   - Which information is available? Which needs to be actively acquired? How can you do that?   - Which information is missing?   - Which tool is useful, which not so?   - What are the problems that can occur when setting the parameter?
Do you take these decisions on your own or with others? With whom?
What are the most complicated decisions that you have to take in your position?
What happens with the production order, once it is issued?   - Who gets it next? And after that? What do these persons do?   - What are the problems that they are facing?   - How do your decisions affect these problems?
When do you deal with the production order again? Are you informed about the further activities related to your order?
How can you evaluate if your decisions concerning a certain production order were good?
What are - from your perspective - the biggest weaknesses in the production planning process? How could the production planning process be optimized with consideration of the goals set by management?

Qualitative research in industry can be very difficult to conduct in a collaborative manner. Since there have usually been some business process redesign projects in the past, often led by consultancy firms, and resulting in job losses due to rationalization, outsourcing or automation, the personnel can be very suspicious towards any kind of external consultants, operations analysts etc. Knowing that, we provided all the involved personnel with a presentation of our institution, an introduction to our project and the objectives of our actual research before commencing the field work. We openly told them that we are interested in how ERP systems are affecting and modifying control on different levels of the organization, also on the shop floor, and how the company could possibly optimize its overall *control capacity*. Nevertheless, and despite of these preventive measures, our position as external investigators might have influenced the responses of the interviewees accordingly.

Now I have arrived at a very critical point of my report. It is not only the people I have been working with that have certain preconditions towards this work, it is also myself that brings certain expectations into the account. For a start, there is my discipline, psychology, and a specific school of thought in work psychology. There is my personal preference for critical and qualitative work. There is a sympathy with ordinary workers and an expectation to find evidence for their everyday contribution to the success of a company, that is usually blinded out. All this certainly has had some influence on my work and needs consideration in the following sections of my report (cf. Parker, 2005b, p. 33f).

Generally, what can be said about my own approach to research and reporting, is that I am very much interested in a 'radical' way of scientific work, including reflection about my own role and limitations. Key issues when writing reports, in my perspective, are also to challenge normative views through reflection of one's own perspective and position, to avoid reductionism and to carefully 'situating' interpretations while also considering those who are affected by the results of my research work (Clarke, 2005, p. xxviiif; Hofkirchner, 2010, p. 14; Parker, 2005b, p. 147f, 2007; Potter, 1996).

5.2 Analysis of planning and scheduling structures and functions

After the first interviews and discussions with senior management we decided to use one of the three business units as our case, hereby delimiting the scope of our study to a part of the company that is of bigger concern than the others and is considerably more complex.

5.2.1 Business unit characteristics

The unit produces 5 million mechanical clock units (movements) per year, which equals 16'000 per day. There are 650 actual variants of the standard products that can be produced. A central sales unit is responsible for customer orders. The planning and scheduling of the production is done centrally in coordination with the sales unit. Six factories are involved in the production process.

Around 35 planners are responsible for the production planning and scheduling, plus a handful more in the factories as local coordinators and shop floor schedulers. The planners used to be organized according to production technologies, but nowadays are responsible for a specific product family and are therefore working across all technologies in the production process. This results in more complexity for the planners that are on-site in the factories (production schedulers and dispatchers). The planners' performance is measured according to delivery timeliness and stock capital.

5.2.2 Orders and planning workflow

The factories and production units do not have their own production planning, but only do work order dispatching. Purchasing of raw materials is done centrally. External production orders and other purchasing is done by the central planners. There are three types of orders: Work orders, purchase orders and subcontractor orders (e.g. for assembly).

In the largest and most important of the factories involved, there is a local dispatcher (called '*Disponent*'), two shop floor supervisors ('*Meister*'), and 5 to 10 workers per supervisor. These persons control and manage 200 machines (lathes) on one shop floor. Planners are in contact with several sales persons and also several dispatchers on the other hand. A dispatcher has to deal with different planners, as well as with different departments within the factory.

The ERP system is planning without consideration of the capacities of single machines. In general, the capacity is based on groups of machines, with the exception of some bottle-necks in the production process. It is the job of the dispatcher to plan which work order will be fulfilled on which individual machine. Until quite recently, some parts would leave a factory unfinished, only to be finished in another factory. The planners were referring to this as 'parts tourism'. This is no longer the case, all parts that are leaving a factory are ready for assembly. Some factories can produce the same parts, as a result.

The volume of work orders has increased substantially in the past three years. Investments have only been made with a lot of delay. Management encourages flexibility, but on-time-delivery has declined from 80% to only 40% of all work orders.

Production time of the overall supply chain is approximately 9 months. Sales orders are fulfilled within 10-12 weeks, even tailored variants according to customer needs. The sales department communicates a fixed delivery date, which creates a lot of pressure on the production side. In order to be able to meet customer demands in this short time frame, production managers need to statistically predict the demand of basic parts and sub-assembly groups (*'fournitures'*). There are so called annual forecasts (*'prévisions'*) which are updated monthly for a time frame of 18 months. These are specifying which standard assembly groups will probably be needed.

From the sales forecast to the actual confirmation of the delivery date to the customer, the workflow of order processing is shown in table 9, as a 'swim lane' diagram.

Table 9: Swim lane diagram for order workflow in business unit

Role / Process	Fore-cast	Program planning	Order planning, dispatching	Pro-duction sche-duling	Assembly and delivery plan
Sales	1. Sales forecast	3. Entry of customer orders, correction of forecasts in ERP			12. Confirmation of delivery date
Central planning department (order preparation)	2. Estimation of model variants; data entry in ERP system	[creation of new variants in ERP, identification, parameters etc.]			
Central planning department (product family planner)		4. Release of production orders on assembly-group level (monthly)			11. Monitoring work progress, scheduling assembly, communication of assembly dates to sales
Production unit planning		5. Release of production orders on part level (planning for 6 months on average)			

Dispatching		6. Confirmation of production order, sequencing and set-up planning	10. Decision on production stop or prolongation of production order (follow-up order)
Worker			9. Feedback on work progress
Shop floor supervisor		7. Set-up orders to individual workers (qualifications matter)	
Technical assistant			8. Realization of set-up plan

5.2.3 Problem areas and potential conflicts

There is a classical conflict for the planners concerning stock capital and cost of machine set-up. In some cases it can take one week to set up a machine. Little stock levels require smaller batches and therefore more set-ups. As opposed to that, high productivity is depending on bigger batches to save costs for labor intensive set-up procedures.

Due to the above mentioned low delivery reliability of only about 40% of all work orders, a working group had been established, lead by the CEO of the company, which also served as the business unit leader at the time. They were meeting three times a week, coordinating and improving the efforts to solve the problems that were involved. In the group were three factory managers, the chief financial officer and the head of the engineering department. The chief quality manager of the company as well as the leader of the central production planning department were invited whenever needed.

Our initial discussions and the subsequent interviews revealed a quite large gap between the realities of planners in the central planning department, the engineers in the IT department and the shop floor dispatchers and supervisors. Their belief in the capabilities and limitations of the technologies and planning algorithms involved were diverging to some extent. An analysis of the decision processes at the various levels in the planning hierarchy made clear that *informal* solutions were chosen under specific circumstances, and the motivations that stood behind that choice were masked using the information system as an organizational *façade* (cf. Abrahamson & Baumard, 2008).

But, neither the theoretical fundament of the M-T-O approach taken nor the available analysis methodology would allow for an in-depth consideration of these forces that would potentially help to explain and possibly also predict planning and scheduling behavior in specific situations. However, an explanation is needed to improve socio-technical design and change management in organizations who are struggling with the implementation of new, and highly networked, information technologies. What is lacking is a powerful and compatible approach to the description of the social 'sub-system' (cf. Keating, Fernandez, Jacobs & Kauffmann, 2001).

5.3 Socio-cognitive discourse analysis and social modeling with *i**

Following these observations in the field, a social modeling approach was used in order to visualize intentional structures held by various groups of actors. This analysis of local rationalities or group discourses using socio-cognitive discourse analysis provided useful in the description of underlying motivations and goals (Gasser, Shepherd & Gärtner, 2010).

In order to achieve that, we developed a methodology based on van Dijk's theoretical perspective on discourse and society that is generally described as *critical discourse studies*, and that could be further specified as *socio-cognitive discourse analysis* in its conceptual orientation (van Dijk, 2001; 2008a, 2008b). The cognitive aspect is necessary to explain how global and local discourses are formed, transformed and reproduced through language and practice. Furthermore, we focus on the particular modes of representation of social practices (Potter, 1996), especially recontextualizations and transformations of these practices (van Leeuwen, 2008a). The method we employed for our data analysis is deriving directly from these methodological traditions.

In our approach we have further employed a pluralistic and heterogeneous view of organizations as complex socio-technical networks. We understand local rationalities and social practices as culturally and historically configured 'mental models' that constitute ways of *knowing* and *doing*. The intentionality behind these social cognitions is inscribed in *context models* that are relevant for the particular social practice. In an attempt to provide a more formal representation of locally enacted intentionality, I propose to use the i^* modeling framework (Yu, 1997; 2009).

According to Yu, all actors (as autonomous entities) should be appreciated in their intentionality when attempting to introduce new technology into an organization. He builds his framework around the concept of intentional actors that are therefore "(...) viewed as having intentional properties such as goals, beliefs, abilities, and commitments (1997, p. 227)". Actors hereby can be human or non-human, or mixed. This approach fits nicely to the analysis of local rationalities I have been advocating so far. Furthermore, according to Yu, "actors depend on each other for goals to be achieved, tasks to be performed, and resources to be furnished. By depending on others, an actor may be able to achieve goals that are difficult or impossible to achieve on its own (1997, p. 227)". We propose the analysis of organizational discourse to identify such local intentionalities and dependencies, which then can be formally described using the i^* modeling framework.

The data analysis consisted of a set of discourse-analytic steps (cf. Figure 12), which were effected with the support of an open-source qualitative research software tool (TAMS Analyzer[11] version 4.45). In a first step, an abstract sequence of topics (the text's 'superstructure') was established. Then linguistic markers related to discourses were

---

[11] http://tamsys.sourceforge.net/

identified. In a third step, the focus was laid on transformations of social practices. The analysis of the material provided evidence for three distinct local rationalities or discourses and connected mental models, including distinctive intentional properties such as goals, beliefs, abilities and commitments which were put together in a fourth step.

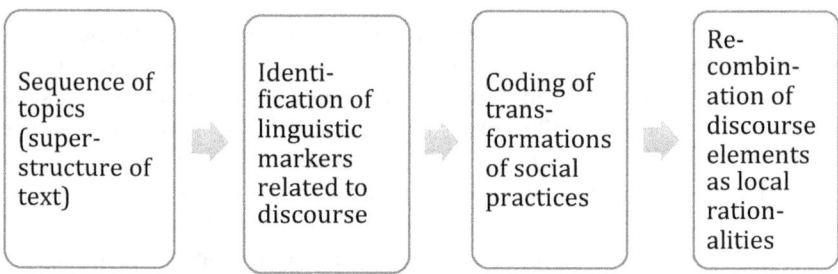

Figure 12: Overview of analytic steps in study one

5.3.1 Description of analytical method

In the following subchapters I will provide a detailed description of the method that we applied to the interview data gathered in this study. A description of the raw data is appended to this thesis, as well as some reports generated using the TAMS Analyzer software to document the coding process and its intermediate results.

5.3.1.1 Sequence of topics / superstructure

To mark and codify the sequential structure of topics that were mentioned during the interviews, and to facilitate orientation within the texts, a set of codes was used. Table 10 is providing an overview of these codes.

5.3.1.2 Identification of linguistic markers for discourse

According to van Dijk (1995; 1997; 2008a), there is a number of linguistic markers for discourse that can be identified. In his conceptualization of discourse analysis, there is a strong focus on ideology and how ideological meanings are expressed and communicated. Therefore, among the countless characteristics of

Study one 133

language and other communications, there is a choice of 'preferential' discourse structures that are used for the analysis.

Table 10: Codes used to establish topical superstructure

Code name	Description
politisch	Political processes, e.g. allusions to influence and power within the organization, decisions without consent and the like.
strategisch	Strategic vision or outlook, strategic decisions by management or other measures related to the organizations development.
administrativ	Administrative processes, e.g. description of order flow or other kind of business process.
praktisch_konkret	Problem or event that has been observed or experienced, description of actual work practice etc.
technisch	Description of technical processes.
sozial	Social processes, including conflicts and diverging thoughts or attitudes.

According to van Dijk, "we need to attend primarily to those properties of discourse that express or signal the opinions, perspective, position, interests or other properties of groups. This is especially the case when there is a conflict of interest, that is, when events may be seen, interpreted or evaluated in different, possibly opposed ways (1995, p. 22)". These selected linguistic features – as 'markers' for discourse – are listed and described in Table 11 below.

Table 11: Codes for linguistic markers of discourse (cf. van Dijk, 1995, pp. 22-32)

Code	Description
oberfl	Surface structures: For example special stress or volume to emphasize or attract attention, the use of irony, of (or lack of) politeness, of accent or dialect.

syntax	Syntactic sentence structures: Active or passive 'voice', word order, agency, complexity of sentences etc.
lex	Lexical choice ('lexicalization'), for example the use of discriminative, racist or sexist expressions, of euphemisms ('surgical strike') or suggestive categorization ('illegal alien' instead of 'undocumented immigrant').
semantik> lokal	Ideologically controlled representation of situations: e.g. biased or distorted attributions to others, positive self-representation of in-groups, use of implicit knowledge assumed to be available by listeners, generality and specificity, impression management ('apparent denial' or 'blame transfer'), de-emphasizing of social inequality.
semantik> global	Topical choice of information: E.g. focusing on one particular theory or explanation for an event, talking of crimes committed by one group (immigrants, minorities) but not others (managers or politicians).
struktur	Schematic structures, topics and their relation to each other: 'Upgrading' or 'downgrading' of information through variation of relevance and importance. Strategic argumentation that leaves by making self-serving arguments more prominent and explicit.
rhetorik	Specific rhetorical structures such as metaphors, understatement, litotes (double negation), exaggeration (hyperbole), euphemisms and mitigation ('collateral damage' instead of 'civil casualties') and repetitions.
pragmatik	Speech acts such as threats and commands, giving advice (without being asked), assertions, impression management, lack of respect, rudeness and other forms of impoliteness.
interaktion	Discursive interaction in conversations: Opening and closing dialogues, turn-management and interruptions, initiation, change and closure of topics, style selection and variation ('breaches' of etiquette) etc.

5.3.1.3 Coding of transformations of social practice

According to van Leeuwen "all texts, all representations of the world and what is going on in it, however abstract, should be interpreted as representations of social practices (2008a, p. 5)". Social practices are socially regulated ways of doing things. They can be regulated or controlled in various ways, through strict prescription, through traditions, through the influence of charismatic role models or experts, through the constraints of technological resources that are used etc. (van Leeuwen, 2008a, p. 6f). In this context it is interesting to recall that Grint and Woolgar consider all technology as 'text' (Grint & Woolgar, 1997). This is in accordance with Rammert, who states that besides describing the 'enrolments' of actors in technological systems or networks, the analysis should include the interactions and relations that lead to these enrolments and how they are changed over time (Rammert, 2007, p. 87).

Texts in various forms, including technology, are representations of social practices. But these social practices are transformed and recontextualized. The recontextualization (in itself a social practice) always takes place in linguistic and/or other semiotic activities, which could be called a 'genre' (van Leeuwen, 2008a, p. 12). It makes the recontextualized social practice explicit to a greater or lesser degree, and it also makes it pass through the filter of the practices in which it is inserted. This process is rarely transparent for the participants since it is usually embedded in common sense thinking, habitual conduct and other tacit know-how. Recontextualization is recursive, it can happen over and over again and thus removing more and more features of the original social practice (van Leeuwen, 2008a, p. 13f).

Van Leeuwen proposes basic transformation mechanisms that can be identified and analyzed in his approach to discourse analysis. The most basic mechanism is substitution. Participants and actions can be either particularized or nominated, herby made more specific, or they can be generalized and aggregated, making them more abstract, vague and 'obscure'. Another mechanism is deletion. Through deletion, some elements of a social practice are omitted. Rearrangements are changes in temporal order or the scattering of elements during the recontextualizing practice, e.g. in a journalistic article. Additions can be formed through repetition, subjective reactions, underlying purposes, justification or legitimations as well as evaluations of actions. All of these mechanisms of transformation are taken to be indicators for discursive practice embedded in the process of recontextualization (van Leeuwen, 2008b). Table 12 is listing the codes we used in the third step of the analysis of our interview material.

Table 12: Codes for transformations of social practice (van Leeuwen, 2008a)

Code name	Description
substitution	Substitutions: The substitution of elements of the actual social practice with semiotic elements (particularization, nomination, generalization, aggregation, objectification, spatialization).
deletion	Deletions: Deletion of elements of social practice.
rearrangement	Rearrangements: Elements of social practice may be rearranged in sequence or scattered through the text.
addition	Additions (repetitions, subjective reactions, purposes, legitimization, evaluation).

5.3.1.4 Recombination of discourse elements as local rationalities

Table 13 is showing the coding statistics that resulted from the coding process.

Table 14 is showing an overview of the coding in relation to the work area of the participant in the study. In total, 403 passages were coded in the 10 interviews that were analyzed (for an overview of interview partners and duration see Table 7 above).

Using the software tool for qualitative analysis, reports were generated to compile sets of quotes related to a work domain and which reveal relevant passages, mainly those with linguistic markers for discourse and transformations of social practices. These reports, one for each work domain, can be found in the annex.

In the following three sub-chapters I am describing the local rationalities that were distilled in our discourse analysis. Quotes from the interviews are within quotation marks, they were translated to English by the author. The interview number is indicated in square

Study one

brackets for reference. The fourth sub-chapter is then dedicated to an *i** representation of these findings.

Table 13: Coding statistics according to code groups

Code group	Code	Number of passages with this code	Number of sources with passages having this code
Superstructure	politisch	23	9
	strategisch	96	8
	administrativ	33	9
	praktisch_konkret	35	7
	technisch	79	6
	sozial	87	9
Linguistic markers for discourse	oberfl	0	0
	syntax	0	0
	lex	26	7
	semantik>lokal	10	5
	semantik>global	0	0
	struktur	0	0
	rhetorik	5	3
	pragmatik	4	3
	interaktion	0	0
Transformations of social practice	substitution	2	1
	deletion	3	3
	rearrangement	0	0
	addition	0	0

Table 14: Coding statistics according to work area and code group (not occuring codes omitted)

Work domain	Superstructure (353)	Linguistic markers (45)	Transformations (5)
Management	politisch (1) strategisch (13) administrativ (4) praktisch_konkret (3) technisch (9) sozial (12)	lex (7) pragmatik (1)	deletion (1)
Central production planning	politisch (9) strategisch (20) administrativ (16) praktisch_konkret (9) technisch (17) sozial (24)	lex (4) semantik> lokal (3) rhetorik (2)	substitution (2)
Shop floor scheduling	politisch (13) strategisch (63) administrativ (13) praktisch_konkret (23) technisch (53) sozial (51)	lex (15) semantik> lokal (7) rhetorik (3) pragmatik (3)	deletion (2)

5.3.2 Management: ERP as an instrument for flexible manufacturing

The material on this level of the organization reflects a general discourse on flexible manufacturing. The goals are to achieve more flexibility in terms of product customization because of market demand as well as a 'lean' production structure with low stocks and little 'waste'. To reach these goals, efforts are under way to further reduce technology-oriented manufacturing in favor of product-oriented "verticalized production [3]". In the interviews, managers are using

Study one 139

terms like "rules of the game [1]", "legislative function [1]", or "standardization of processes [1]" that indicate a rather essentialist or technicist perspective on technology. According to them, there are many "wishes [1]" within the organization, and someone has to use the "competence [1]" to impose a certain design. But according to one manager, it should not be enforced without consent: "I usually try to persuade people [1]." Overall, the management revealed an instrumental perspective on the ERP issue.

Concerning intended functionalities regarding production planning, the IT manager responsible for the ERP implementation project stated that he wants to include a "detail planning module [1]" for the shop floor. Hereby he was indicating that he believes it will be possible to implement the system in a way that reflects the realities and contingencies on the shop floor in sufficient detail and quality to allow for such 'detail planning'. He stresses that on the shop floor, there is "currently no support [1]" at all, hereby omitting that the shop floor planners have self-made tools for production planning that seem to be sufficiently functional for them. This omission (deletion) of an existing social practice again indicates an instrumental managerial perspective: The right technology will solve all problems - regardless of current practices - under the implicit commitment to substantial organizational change.

5.3.3 Central production planning: ERP as an administrative tool

The planners in the central planning department would like to have reliable partners in the production centers to facilitate their work of coordinating the production and assembly of ordered products. They would like to further establish a "pull principle [9]", which means that the delivery of ordered parts should happen without further intervention on their side. In their perspective, the factories are more or less like "individual companies [9]" that take orders from them and have to make sure that the delivery follows on time. They are arguing for a further "verticalization [10]" of production, mainly because of the lack of transparency (the backlog list is like a incomprehensible "book [9]"). They also mistrust the production units in terms of their estimated production capacity. One of them is ridiculing a factory manager by quoting him as saying "oh, I think we need a new automaton [10]", meaning that he thinks that maybe there is not enough capacity, but without actually knowing that for sure.

There is some criticism of internal product development and marketing practices: "Marketing already done, that is deadly. Such things [new product development] must be clear from the beginning and realistic. Not because someone wants to be satisfied in short term and then the whole chain gets problems [2]." But when it comes to sales and customers, the criticism is rather weak: "When the customer is counting on the parts, has already ordered everything for his market, and maybe he has already been marketing the product, then there is big trouble [2]." Instead, the planners are using strategies to keep a clean record by getting things "in written form [2]", and therefore are indicating that they prefer to stay neutral in this power play. Others decide when it gets critical: "Marketing does that, that is very political [2]". And they avoid blaming from this side by making sure their part is done: "Not that they come and tell me I did not order the parts [2]." The customer is the ultimate justification for prioritizing: "The customer wants the first piece immediately, so that he can start to work [with it] [2]." The shop floor scheduler is hereby seen as a subordinate instance that "has to take care for the data [2]". At the same time, the planner is complaining about the lack of feedback from the factory, saying that "they are just eating everything, they have almost capitulated [2]". Interestingly, he compares the 'capitulation' of the shop floor planners with the 'capitulation' of the central planners vis-à-vis marketing: "...then I call [the marketing], and ask them why, and they tell me, yes, that is political, we can't do it differently [2]."

For the central planning department, the ERP system is mainly an administrative tool to track (mostly unquestioned) marketing demands, to justify own planning activities in case of trouble and to check on production progress. Its failures are attributed to an insufficient production structure, i.e. incomplete "verticalization [9]", and a lack of responsibility on the lower levels of the planning hierarchy. As a planner frames his discontent with the shop floor schedulers: "They do not know what the consequences [for the assembly plant and the customers] are [9]". The planners' general perspective on ERP as a technology could be described as purely technicist.

5.3.4 Shop floor scheduling: ERP as a reality-ignoring system of justification

From a shop floor perspective, the ERP system is perceived very critically. According to the schedulers, it does not represent "reality [6]". Central product planners are using it to "safeguard [6]" themselves against management, and everybody is practicing "system cosmetics [6]" to mask failures: "It is all a lie [6]". No one in the planning levels above wants to "put his head on the block [6]" for making a decision between equally urgent orders. Due to this unwillingness, the scheduling decisions are made locally, since it is impossible to work according to the due dates, because "we have to work with priorities, with capacities [6]". When they fail to deliver in time, the system is used to blame it on them mostly. But the shop floor schedulers are questioning such argumentation: "The question is, is it our backlog or the backlog of the whole company? (...) we knew that we couldn't make it from the beginning [6]." They double check system contents like orders and stock levels constantly, to establish a "truth [6]" concerning the most pressing needs: "It is believing, disbelieving, trust, distrust [6]."

The ERP system does not play a role in the most pressing planning problems of the shop floor, unexpected urgent customer orders and technical changes that lead to rework. But these often lead to further delays and backlogs not represented in the system. The upcoming implementation of a new ERP system is seen in the context of general management pressure towards smaller batch sizes and a more flexible production, because "they would like to tune that [10]". There is also a fear of an incomplete further "verticalization [8]", leaving shop floor schedulers responsible for certain product groups, but without dedicated production resources. One of the schedulers believes the result will be that "we will seize each other by the neck [6]".

Without being able to suggest alternatives, the local schedulers remain skeptical towards these management initiatives. In their view, operational uncertainties, "technical problems of feasibility [4]" as well as a necessary "base load [5]" to retain and develop technical know-how within the factory delimit such efforts substantially. As one scheduler stated, "[the central planners] have to adapt to the production [6]". They utilize a local rationality based on a notion what is "reasonable [10]" in terms of resource allocation depending on job difficulty, set-up times, stock levels and batch sizes. Too little batch sizes are considered as "indecent [10]" to work with since they lead to less satisfaction and more stress on the side of the operators. As one of the shop floor schedulers put it: "This is no way of working [10]". Generally there is a strong affirmation of social coherence in the factory. In sum, the shop floor schedulers formulate a critical view of enterprise information systems that are detached from production realities and an intuitive awareness of social implications of ERP deployment and configuration strategies.

5.3.5 Distributed intentionality and strategic dependencies

The description of local rationalities as elaborated above subsequently allowed for a formulation of *strategic dependencies* as proposed by the *i** framework (Yu, 1995, 2009). Yu's Strategic Dependency (SD) model describes a process in terms of intentional dependency relationships among agents. Agents depend on each other for *goals* to be achieved, *tasks* to be performed, and *resources* to be furnished. Agents are intentional in that they have desires and objectives, and strategic in that they are concerned about opportunities and vulnerabilities (cf. Yu, 1995, p. ii).

In the Strategic Dependency model, we assume that participants in software processes are organizational actors who need to cope with problems cooperatively and on an on-going basis. How actors make use of, and constrain, each others' problem solving activity is therefore an important aspect of a software process that needs to be modeled and reasoned about (cf. Yu, 1995, p. 97).

The strategic dependencies we have identified from the analysis of work domain specific discursive elements are shown in Table 15. These results can be made more accessible using a graphical notation proposed by Yu (Yu, 1995). The resulting model is shown in Figure 13.

Table 15: Strategic dependencies distilled from discursive elements

Actor	Discursive elements	Goal, task, resource dependencies
Manager	*Lean; waste; verticalized production; rules of the game; legislative function; standardization of processes; wishes; competence; I usually try to persuade people; fine planning module; fine planning; currently no support.*	Wants to achieve flexibility (goal), wants to set rules and standards (resource), wants to be assured (soft goal), wants to get consent (soft goal).
Product Planner	*Pull principle; individual companies; verticalization; book; oh, I think we need a new automaton; marketing already done, that is deadly. Such things [new product development] must be clear from the beginning and realistic. Not because someone wants to be satisfied in short term and then the whole chain gets problems; When the customer is counting on the parts, has already ordered everything for his market, and maybe he has already been marketing the product, then there is big trouble; in written form; Marketing does that, that is very political; Not that they come and tell me I did not order the parts; The customer wants the first piece immediately, so that he can start to work [with it]; has to take care for the data; they are just eating everything, they have almost capitulated; capitulation; ...then I*	Wants to keep 'clean' (goal), wants to justify activities (resource), wants to check progress (resource), needs information from marketing (resource), wants processes to be managed (resource), wants to provide feedback (resource), wants to avoid prioritization (soft goal), wants to delegate responsibility (soft goal)

	call [the marketing], and ask them why, and they tell me, yes, that is political, we can't do it differently; Verticalization; They do not know what the consequences [for the assembly plant and the customers] are.	
Shop floor scheduler	Reality; safeguard; system cosmetics; It is all a lie; put his head on the block; we have to work with priorities, with capacities; The question is, is it our backlog or the backlog of the whole company? (...) we knew that we couldn't make it from the beginning; truth; It is believing, disbelieving, trust, distrust; they would like to tune that; verticalization; we will seize each other by the neck; technical problems of feasibility; [the central planners] have to adapt to the production; reasonable; indecent; This is no way of working.	Wants a fairly reasonable schedule (goal), wants to check the 'truth' about orders (resource), wants others to acknowledge 'reality' in production (soft goal), wants decent work (soft goal), wants to protect social coherence in the factory (soft goal)
ERP software	[dependencies inferred from the above findings]	Depends on order releases (task) and progress reports (task)

Study one 145

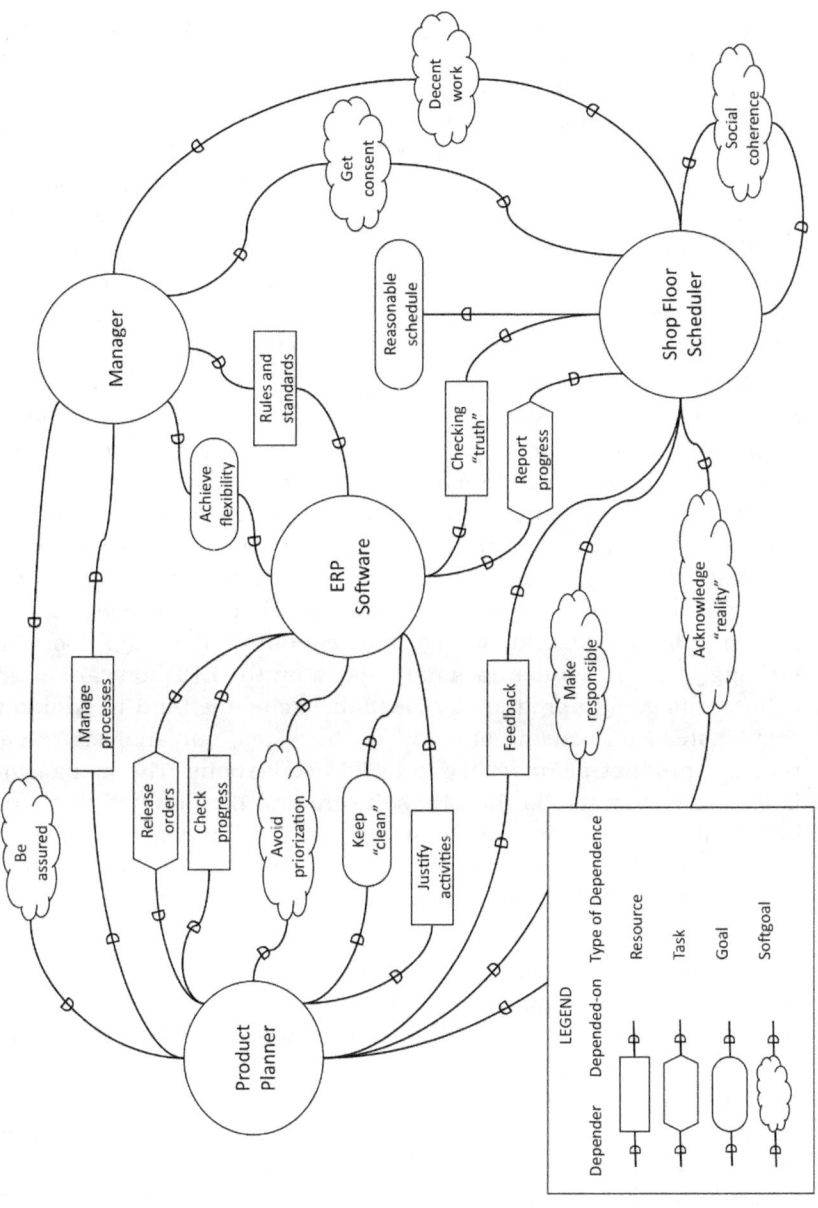

Figure 13: Strategic dependency model for production planning and scheduling

Actors are represented in circular containers. Four different types of dependencies are shown, namely resource, task, goal and soft goal dependencies between actors. To keep the model manageable, marketing and sales as well as shop floor personnel have been left out. Of course there are some dependencies as well. Product planning depends very much on reliable forecasts and timely order placement. On the other hand, scheduling is depending on compliance and on time dispatching and execution on the shop floor. Although it only represents a small part of the overall organization or even the planning hierarchy, we believe that the model does not substantially suffer in its descriptive power from these omissions.

The model shows the principle of delegation between planning and scheduling. The ERP software is depending on tasks fulfillment on both sides. Orders need to be released, and progress needs to be reported. In the other direction, the product planner seems to be much more dependent on the ERP software than the shop floor scheduler. In addition to that, he is also dependent on feedback from the scheduler. He would therefore like to make the scheduler more responsible, whereas in the opposite direction, the scheduler would like the planner to acknowledge the realities of the shop floor. Most strikingly, the scheduler does not depend on the ERP software at all to achieve his goal to produce a reasonable schedule. And in addition to that, instead of focusing on avoiding backlogs, our analysis revealed that the product planners try to avoid accusations. The management is relying very much on the ERP software and the product planners. It is willing to invest in participatory, consent-building activities, but only to a certain extent. The schedulers - being close to the shop floor - are addressing the management implicitly to ensure a decent work environment in the factory.

The substantial amount of resources needed to implement such a software must be justified through massive gains in productivity and therefore profit. It seems that managers willing to buy this product share a belief that they can achieve a reduction of 'waste' (in terms of lean management), if only they see clearly what is going on in their company. There is an element of totalitarianism in this: With detailed and actual information, one can eliminate all 'slack' and inefficiency single handed. This is in perfect correspondence to what Streatfield has described as the mainstream perspective on management practice: "(...) managerial action forms the organization and its movement. This movement requires the human agents of the system to act in conformity and sustain consensus. The goal and the path are largely known and they are formed by management intention so that the movement of the organization is stable, regular, predictable and in principle certain. The movement expresses the identity of the organization as continuity, which implies the habitual movement of culturally determined behavior in which people share the same values. The resulting clarity [sic!] is secured by conscious managerial decisions having the characteristics of formality and legitimacy (2001, p. 126)".

According to Streatfield, mainstream managerial thinking focuses its attention on the movement of the whole system, hereby aiming at 'unfolding' the future in a continuous process that is informed more by the past than the present. It requires an individual objective observer with a lot of hierarchical power (Streatfield, 2001, p. 131). However, this perspective and related managerial practices have far reaching consequences on organizations as work environments. It requires the other human agents to 'act in conformity' and to share culture and values. Here, the gap between mainstream managerial thinking and the reality in most of today's work places becomes obvious: There are very few globalized organizations that fulfill these preconditions.

5.4 Intermediate discussion

In case study one, we have analyzed planning and scheduling structures, their intentional and discursive context as well as strategic dependencies between the actors involved.

In Chapter 4.5, we have identified that there are three things needed for our theory/methods package:
1. A critical realist *local / scientific ontology* for Human Factors as a template or background, covering both technological and social spheres,
2. a *set of adapted / extended analytical methods* to develop local theories with the help of retroduction and retrodiction and
3. a *self-critical, incremental research process* that allows for validation and evaluation of the explanations and theories that we develop.

5.4.1 Social discourse analysis, ontology and retroduction

It almost seems that - when we think of mutual strategic dependencies such as in our analysis of distributed intentionality - *power relations* do not exist within the organization. The world seems very 'flat' indeed (cf. Reed, 1997). But of course this is not the case: It is generally agreed that there is a power distribution within organizations that in most instances allocates much more power on the side of management. Nevertheless it is true that every single manager is very much dependent on delegation of tasks, the consent and the authorization of others. Clegg and Wilson thus formulate the paradoxical nature of power: "The power of an agency is increased (...) by delegating authority; the delegation of authority can only proceed by rules; rules necessarily entail discretion and discretion potentially empowers delegates (1991, p. 245)." Yu is well aware of some of the implications of mutual dependencies without referring to theoretical approaches to power. But he makes a point that is very relevant to our discussion of information technology in organizations by saying that "an actor becomes vulnerable if the depended-on actors do not deliver. Actors are strategic in the sense that they are concerned about opportunities and vulnerabilities, and seek rearrangements of their environments that would better serve their interests (Yu, 1997, p. 227)". Therefore we could assume that - given the power relations in organizations - more powerful but dependent actors might be inclined to seek such 'rearrangements' when implementing new technology. More explicitly, management might promote ERP in its own interest as an instrument of control.

Although this seems compelling, one should avoid simplistic and one-dimensional perspectives on organizational power relations, especially when thinking about technological change such as the implementation of an ERP system. I believe that it is helpful to let go of a zero-sum concept of power relations - i.e. management rules, workforce resists - and to adopt a dynamic, multi-centered and heterogeneous view on discursively established structures that are empowering organizations. Or, as Clegg and Wilson put it, "it is with (...) the institutional frameworks of available knowledge in and around organizations that the multiplicity of potential centers of power within organizations might seek to enhance their strategicality and thus their power. Within the organizational arena agents with varying strategies are seen to struggle to constitute the capacities of the organization in policy terms which represent their conceptions of their interests (1991, p. 246)".

In any larger corporation or organization, managers are working in networks. They are forced to engage in mutual negotiations of dependencies and alliances. From such a perspective, management skills and competencies lie in how effectively managers participate in those processes. It is asking what competent managers actually *do* to live effectively in the paradox of organizing. According to Streatfield (2001, p. 128), what they actually do is continue to interact communicatively, especially through the means of conversation, in spite of not knowing and not being simply 'in control'.

But one has to be cautious about ideological influences when reflecting upon managerial practices. Mainstream Fordist thinking about what managers do and should do is not only affecting managers themselves. External influences on organizations, for example in the from of technology itself, or through its vendors or affiliated consultants, might also be at work that affect organizational politics (Benders et al., 2006; Koch & Buhl, 2001). The believe that the right technology will allow a manager to 'be in control' at any time stands in conflict to the inherent necessities and uncertainties of the actual work process that involves people with a wide variety of cultural and personal backgrounds.

These tensions have been described by other authors as well. For example, it is found that management often uses game metaphors to legitimate possible sanctions (van Dijk, 2008b, p. 52). The main interest seems to be the establishment of a rule-based institutionalized structure[12] that *constrains* different actors in the organization (Gosain, 2004). Very often, the main driving forces behind management initiatives to implement ERP systems are integration and dis-embedding of work practices through a wide-ranging standardization of processes (Hanseth et al., 2001).

5.4.2 Retrodiction, model building and critique

The identification of explicit and implicit goals of normative circles (groups) or centers of power within the organization as well as their tensions, potentials, mutual dependencies and means of influence

---

[12] These structures can also be conceptualized as *discursive norm circles* (Elder-Vass, 2012a).

now must lead to retrodiction and model building to explain actual events. In this case study, we have come closer to a more 'in-depth' analysis *why* ERP implementation projects fail completely or are generating very high costs and enormous delays.

Nandhakumar, Rossi and Talvinen are providing an example case study on a company called EURMOBIL that produced comparable results with a different methodology (2005). In their study, they found that "both affordance of the system and social structure allowed or restricted the managers' intentions with respect to ERP implementation. When there were restrictions they often resorted to revising development plans or rescheduled events; changing existing or new technological components; changing organizational processes; or abandoning their plans. At EURMOBIL managers therefore tried hard to configure the ERP system to gain control over its functioning. The Corporate ERP however never emerged; instead a fragmented, new Corporate ERP emerged out of these new components. (...) The planned global 'unifying model' of the ERP solution never materialised along with several planned components but many of the realized ERP components were unintended and emergent. A consequence of these emerging projects is the technological drift (Nandhakumar et al., 2005, p. 237)". Nandhakumar and his colleagues are proposing an exploratory model to show how drift and control in ERP implementation projects are emerging (cf. Figure 15). The model illustrates how the process of implementation was triggered by designers' and users' intentions (intentionality) in response to internal and external contextual conditions. The perception of what action is possible with the technology (affordance) and the power and cultural settings (social structure) in turn shaped the implementation process through *cycles of interaction* of intentionality, affordance and social structure. On one hand managers attempted to configure the system in ways to achieve control over material phenomena. On the other hand the affordance of the system and the social structure allowed or restricted managers' actions and intentions with respect to ERP implementation. The ERP system implementation therefore 'drifted' as a result of the matching between situated human interventions and the ERP system's properties and functionalities (Nandhakumar et al., 2005, p. 237).

Study one 151

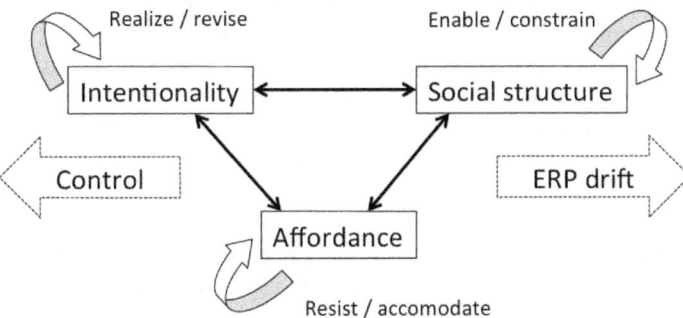

Figure 14: Model of drift and control in ERP implementation (adapted from Nandhakumar et al., 2005, p. 238)

What Nandhakumar and colleagues are lacking is an ontology to allocate and embed their findings in. They stop their analysis at the very moment where retrodiction should begin. In the following, I will try to go at least one step further by referring to a critical realist ontology as sketched out in chapter 4.5 further above: Organizations are structured social groups with emergent causal powers. They depend on normative mechanisms to produce the *role specialization* upon which they depend, but *role-coordinated interaction* between their members (which may include non-human material things) provides a further class of mechanisms, a class that confers non-normative causal powers on the organizations concerned. Authority relations are a variety of role specialization, they confer some part of the power of the organization as a whole on certain persons (as role occupants). The management role includes the development of the role specifications in response to the goals, performance and circumstances of the organization. Technological systems are social structures, too. They embody norms of behavior, degrees of freedom or constraints to agency that have been inscribed to them by their designers on behalf of their customers.

The fundamental issue at hand seems to be what Streatfield has called 'the paradox of control' (Streatfield, 2001). Managers often favor a Neo-Fordist perspective in terms of control of production which becomes evident through their belief in the ERP-supported increase of product innovation and process variability, but without real increase in worker responsibilities (Clegg & Wilson, 1991). In our case, without critically reflecting its own moderating role in the area of tension between marketing and production, central planning takes a neutral brokering perspective within the organization. Standards are in place, processes streamlined. Opposed to that, shop floor scheduling predominantly relates to the community of practice within the factory and local rationalities connected to them. These diverging discourses lead to inconsistencies of intentionality that are harmful for the overall efficiency of the organization. Nauta and Sanders (2001) have found in their study on perceived goal differences between departments that the consequences of goal differences are an increased number of conflicts between departments. An interpretation of our study results also suggests that ERP technology implementation following a predominantly technicist discourse is leading to less inter-departmental communication and coordination which in turn causes an increase in goal differences and conflicts.

In addition to Nandhakumar and colleagues' general findings about ERP implementation, we have been able to differentiate not only the notion of intentionality, but also of social structure. We have described three local rationalities that encompass normative beliefs and dispositions related to a certain practice. These 'position-practices' (Bhaskar, 1979, p. 40; Mutch, 2010, p. 551) or role specializations are (re-)produced by normative mechanisms. At least in the case of (Fordist) mainstream management we have been able to describe one of these mechanisms in more detail. There are possibly others, but our empirical material – stemming from a project that was predominantly focussing at planning and scheduling functions – is not providing any evidence of a special 'workers culture' or related social norms. Nevertheless, such influences probably exist. As in Nandhakumar and colleagues' findings, we would assume a potential for conflict or drift when incompatible cultures of the relevant social groups in an organization meet each other in a complex change project.

Furthermore, the aspect of technological *affordance* has not received its due attention in our study (cf. Hutchby, 2001). There is a possibility that the explanatory power of our methodological approach could profit from a more thorough analysis of the properties of the technology that is about to be implemented. What I have in mind here are not the purely technical aspects nor the functionalities of the software. What needs to be analyzed are the sociomaterial fundaments or 'social institutions' that are built into technology and therefore have 'causal power' – possibly approaching what Orlikowski has called 'sociomaterial configurations' (Orlikowski & Scott, 2008), but with a critical realist ontology in mind. For example, user roles and rights or process work flows could be used by developers to implement social structures within a software product.

A second thought is concerning the subversive nature of all kinds of *'workarounds'*, *'system-cosmetics'* and other local practices that can be observed in these work environments (cf. descriptions of technologies-in-practice in Orlikowski, 2000). Coming from our analysis, I admit that we have not been able to produce good explanations for these phenomena. What purpose exactly do they serve? – One preliminary hypothesis could be that they allow workers to sustain a minimal autonomy in the face of the mighty control regime that governs their time at work. Their fundamental concern is decency of work as well as social coherence in the team or department. A little episode from the factory we have studied might illustrate that: A skilled worker was mandated to set up a lathe for the production of very difficult parts. The parts were so small that the process came very close to the technologically feasible. He managed to set up the machine in a week's time. When he finished, the orders had changed so that he had to set up the machine once more, but for another part, without having produced a single item so far. After the second week, the scheduler came up with yet another schedule. This time, the shop floor supervisors ordered the scheduler to personally explain this to the worker that had already been working two full weeks 'in vain'. The scheduler, who was telling me this story, told me that this was a very troubling moment in his career as a planner – something he certainly is going to avoid by all means in the future.

Both considerations, the 'implicit' features of the technology as well as the 'subversive adaptations' to them need our attention as researchers, if we want to explain all aspects of how and why ERP implementations falter or thrive.

5.4.3 Lessons learned

From case study one, what can be learned for the project of a critical theory/methods package for Human Factors? Our approach has shown that socio-cognitive discourse analysis can be used to describe local rationalities that are at work in the organizational arena, using the pre-implementation stage of an ERP system as an example. We have identified and addressed inconsistencies and tendencies concerning technology implementation by applying early-phase requirements analysis done from a critical, post-essentialist point of view.

Our analysis shows fields of tension that could potentially - at this stage - be modified and/or considered in system design. A moderated participatory pre-implementation dialogue on the meaning and purpose of the technology might then lead to a different deployment strategy and design of the overall ERP system and its functionalities. Such an approach still requires an organizational change process, but it will not be solely driven by technical or other demands created by the ERP software itself. The change process would possibly profit from a more comprehensive analysis and reduce change-related effort and cost (e.g. through a more structured approach to user participation). In this way, I have already reached some practical value with my research. I have been doing so with considerable but – given the size of ERP implementation projects – still reasonable effort.

However, the identified fields of tensions between position-practices and various social and organizational phenomena emerging from them are not yet fully understood and explained. The methodology is stepping short to reveal the full scope of causal powers employed by the different social structures in place. Especially, the question of how ERP systems are dynamically supported and undermined at the same time remains to be addressed. A more dialectical perspective might be of use for this endeavor. But, are we really allowed to ask *why* users do not accept certain socio-technical design propositions (as, for example Liu & Yu, 2004, pp. 6, does)? And if so, do we always get valid answers given that, as researchers, system designers and management consultants, we are possibly not perceived as 'neutral' participants in the process? And what if users are not fully aware of the underlying reasons of their resistance?

What Nandhakumar and colleagues (2005) are *not* explaining is *why* the intended ERP system was not successfully implemented. In the end, there was a plethora of systems in place at EURMOBIL. Who was opposing the system, and why? In his book on the 'Forces of Production', Noble (1984) analyzed in detail how workers opposed to the implementation of computer-controlled mechanical lathes in their workshops. One of their reasons was that these machines would completely change their work, and thus for them the question of technology became an existential issue of *quality of working life* (cf. also Rasmussen et al., 1994, p. 96). In their view, automated lathes would make them dependent on an engineer to do the programming and would reduce their duties to refilling the machines with raw materials and other 'caretaking' chores. This would then lead to a decrease in skills and competencies on the shop floor level and consequently also lead to a smaller salary. Therefore, the conflicts around automation and new technologies in these factories were as much about control and autonomy as they were about decent work and pay.

From a critical or dialectical point of view, every analysis must ask "of every 'thing' or 'event' that we encounter by what process was it constituted and how is it sustained (Harvey, 2004, p. 126)"? One of the questions we must address therefore is how ERP systems have been constituted and how they are sustained. It is only in this way that we can later identify possibilities to transform their biases and side-effects into something of greater value for all participants in a production process. In its essence, this endeavor is what others have termed *critical information systems research*. As Howcroft summarizes, "it questions and deconstructs the taken-for-granted assumptions inherent in the status quo, and interprets the development and adoption of information systems by recourse to a wider social, political, historical, economic, and ideological context. In this respect it takes issue with how information systems can be used to enhance forms of control and domination. It is intended that this process of analysis may help uncover systems of institutional repression and human resistance in order to initiate social change and reform (2009, pp. 392-393)".

However, critical accounts of local rationalities or intentionalities as well as specific practices are not to be reduced to culture or class struggles. Doing so would not do justice to the agency and power of the groups of professionals involved. Neither would a simplistic or reductionistic consideration of the technologies at hand reach the goals of our analytical approach. As Bhaskar aptly formulated some time ago, we need a mediating system, a system that could be called the position-practice system (Bhaskar, 1979, p. 40f).

These considerations are showing the need to address the more fundamental question related to the organizational 'regime of control' that is shaping and controlling ERP system design. I would like to tackle this in the second case study of this thesis by using a different method, based on Archer's *morphogenetic approach* (Archer, 1995). In doing so, I will reconsider our own ideas about structural and operational control capacities (Wäfler et al., 2008). Besides that, I will also relate my discussion to critical realist studies of organizations that explore different forms of power and the longer-term institutional consequences flowing from the generative mechanisms that emerge from them (Reed, 2009). Related to this is theoretical and empirical work on human agency in ERP implementation and use, for example the 'double agency' perspective (Ignatiadis & Nandhakumar, 2006), or the concept of 'enactment' of ERPs (Boudreau & Robey, 2005). Study two is thus aiming at a deeper understanding of these dynamics that form and sustain agents and structures in organizations using ICT for control purposes.

6 Study two

> Technocracy need not impose a specific value-based ideology vulnerable to critique on factual grounds. Rather, it relies on the consensus that emerges spontaneously out of the technical roles and tasks in modern organizations. Controversies are routinely settled by reference to that consensus. Meanwhile, the underlying technical framework is sheltered from challenge (Feenberg, 1999, p. 103).

To explore the complex interplay of various organizational actors to plan and control production processes in manufacturing, a case study approach was used by a team of interdisciplinary researchers coming from Business Management, Information Sciences and Human Factors backgrounds. The unifying concept was that of complexity and the human capability to deal with it when empowered to do so.

6.1 Approach and methods

The study has been conducted in a Swiss manufacturing company using a set of qualitative methods to investigate the planning and scheduling work domain as well as critical control tasks from a Cognitive Work Analysis perspective (cf. Naikar, 2006; Rasmussen, 1986b; Rasmussen et al., 1994; Vicente, 1999). The investigation took place in an interdisciplinary project and was motivated through the need to improve the overall planning performance in the company.

In our case study, we generally followed the five maxims that were defined by Button and Sharrock (2009, pp. 61-68): (1) Keep close to the work, (2) examine the correspondence between the work and the scheme of work, (3) look for troubles great and small, (4) take the lead from those who know the work, and (5) where – in which context – is the work done?

6.1.1 Research hypotheses and topics of interest

Our interdisciplinary approach was inspired by previous projects and case studies that we had done together (cf. Gasser et al., 2007). Our questions and hypotheses were derived from these studies and discussions:

- What are the consequences of operational uncertainties, disruptions and fluctuations (e.g. over- or underload, urgent orders) on the choice of planning strategies?
- What are the 'control modi' that can be distinguished?
- Which control tasks and strategies can be distinguished and to which extent do the software tools in place support them?
- What are the consequences of various 'control states' and control strategies on the collaboration between departments?
- What is the role of ICT hereby?

6.1.2 General objectives and intended outcomes

The general objectives were related to the coping strategies of the planning network when dealing with complexity. Control tasks require access to distributed knowledge and adequately functional decision hierarchies within the organization. Our internal goal was to come closer to a proposition for an alternative planning method that would not be based on ever more sophisticated algorithms but on decision support for planners that takes their human capabilities in account. For our client, we were aiming at 'quick wins' out of the research process, mostly in the form of workshop results to enable internal creativity, inter-departmental communication and solutions.

We expected that the study would lead to various outcomes on different levels. Firstly, on the level of exploration and description, we wanted to establish a complete description of the planning and scheduling work domain (decomposition-abstraction matrix). Second, we were aiming at a catalogue of operational uncertainties that affect the planning process. Thirdly, we wanted to identify decision making processes in connection to operational uncertainties at different work places. Fourth, inter-departmental coordination to solve urgent problems would be described following decision making efforts. And fifth, we were looking for control strategies that fit to certain control situations.

On the level of methodology, we planned to develop further an instrument to identify and describe operational uncertainties as well as a method to do decision probes (a form of semi-structured interviews).

Theoretically, we were interested in more empirical evidence for our model of control capacity (Karltun, von der Weth & Gasser, 2006; Wäfler et al., 2008). The main question in this regard was if our distinction between structural and operational control capability was of practical value for organizational analysis and socio-technical design.

6.1.3 Methods and research plan
Our approach was adapted from Rasmussen and colleagues' Cognitive Work Analysis framework (Rasmussen, 1994; Vicente, 1999). The reasons that lead us to this choice were (1) that the method was developed for complex socio-technical systems, (2) is aiming at an objective analysis of the boundary conditions for behavior, especially control tasks, (3) is hereby successfully avoiding a reductionistic information-processing perspective through its differentiation from normative task analysis. (4) CWA uses the notion of 'strategies' to describe control – or problem solving – behavior in various contexts and situations and it (5) does not restrict itself to one workplace. Finally, it (6) is open for selected additional elements and categories, for example operational uncertainties.

In order to accommodate for our own goals in this case study, CWA was slightly modified and extended. There were two main additions that we developed, one aiming at operational uncertainties and the other at decision making processes within control tasks. Theoretically and terminologically we oriented ourselves towards the traditions of work psychology based in Berlin, Dresden and Zürich.

Our extended and modified version of CWA consisted of 7 steps instead of the original 5 steps. Table 17 is showing an overview of these steps as compared to the original CWA method (Rasmussen et al., 1994; Vicente, 1999). Step one was modified, steps two and three were added and steps 4 to 7 were explicitly focused on planning and scheduling tasks.

6.1.3.1 Step 1: Hierarchical functions analysis
A defined part of the work system, here the domain of production planning and scheduling, is analyzed according to its main functions. Hierarchically, the analyst asks, what function is executed by whom with which inputs and what output. In general, outputs are providing inputs for the next function(s). The function can include a multitude

of inputs and can consist of several tasks. For example, the planning of the repair process needs various inputs (order details, assessment results, work plan, due date etc.). The function itself requires the modification of the standard work plan as well as the reservation of certain critical resources and the dispatching of individual work orders. The result of this analytical step is a set of hierarchically nested input-output diagrams with descriptions of the functions that form the sub-processes of the overall control network.

Table 16: Extended and modified Cognitive Work Analysis method for control tasks

Cognitive Work Analysis	Extended Cognitive Work Analysis for Control Tasks
1. Work Domain Analysis (Abstraction Hierarchy)	1. Hierarchical functions analysis (input-output) 2. Goals and constraints for planning tasks (means-ends) 3. Catalogue of operational uncertainties
2. Control Task Analysis (Decision-Ladder) 3. Strategies Analysis (Information flow)	4. Task analysis of control tasks on the workplace level 5. Decision analysis of control decisions
4. Social Organization and Cooperation Analysis 5. Worker Competencies Analysis	6. Interdependencies of control tasks (information flow, feedback) 7. Used vs. potential control possibilities and capabilities (competencies)

6.1.3.2 Step 2: Goals and constraints for planning tasks

The objective of the next analytical step is to identify constraints for planning tasks. To achieve this, the input-output diagrams elaborated in the first step are used to conduct structured interviews with the persons involved. The main question is targeting means-ends relations (what needs to be done to achieve which goal?). The second, related question is targeting the constraints that have to be accounted for. In

Study two 161

the background, the researchers are constructing a work domain to allocate control tasks in a hierarchical matrix (decomposition-abstraction matrix). The related questions are: Which are the boundary conditions that need to be considered? What control tasks can be derived from them? Who is responsible for the execution of these tasks? Like this, a hierarchy of goals, functions and a decomposition of the work system with the use of means-ends relations can be achieved. This composes the work domain for control activities in planning and scheduling.

6.1.3.3 Step 3: Catalogue of operational uncertainties

Next, the establishment of a catalogue of operational uncertainties allows for the identification of control tasks. Since it is not feasible to ask research participants about the control tasks they are responsible for, this is a way to come up with a list of 'problems' that need to be solved in the work domain – eventually leading to the formulation of control tasks. The methods are semi-structured interviews and diary recordings at the workplace (online questionnaire with daily reminder). The questions to ask are:

- What are operational uncertainties and how often do they occur?
- What is the impact of these uncertainties, i.e. on transparency, predictability, manageability, potential for disruption or damage?
- How do you cope with them?
- Who is informed or otherwise involved?

The main result of this analytical step is a comprehensive list of operational uncertainties in the work domain with a rating of incidence and consequences, i.e. relevance and thus 'control need' of each particular uncertainty.

6.1.3.4 Step 4: Task analysis of control tasks on the workplace level

Control tasks are derived from operational uncertainties that are in some way manageable or controllable. In theory, every uncertainty generates a need for control actions, which in turn is met by a control task that is allocated to a person or team. This step of the analysis is aiming at the description of these control tasks at the workplace level. Its objective is to describe decision latitudes and opportunities for

control within these tasks. The method is semi-structured interviews. Questions to ask are:
- What is your mandate? What goals have been set?
- What are the boundary conditions and constraints for this task?
- What is the decision and action latitude for this task (operational, structural)?
- What are the decisions that need to be taken? How often?
- What are control opportunities that are not exploited? What is desirable from the perspective of the worker?
- Which role do ICT play in this?

As a result, this analytical step produces a detailed description of the control tasks within the work domain. In addition, a list of control task related decisions as well as a list of control opportunities that are currently 'used' as well as 'unused' is compiled. This allows for a first estimation of the fit between control *need* and control *capability* or *capacity* on different levels of the work domain (cf. Karltun et al., 2006; Wäfler et al., 2008).

6.1.3.5   Step 5: Decision analysis of control decisions

The previous step has produced a blueprint for more in-depth workplace observations. The focus hereby is on decision-making behavior. What are the control-related decisions that need to be taken? Under what circumstances? How often? With which tools / aids? Are they distributed or local? Does the decision-maker consciously use any cognitive strategies, rules, heuristics or the like? The method is systematic observation at the workplace, semi-structured interviews and theory-guided analysis, based on the 'decision probe' method developed by Wilson and colleagues (Jackson et al., 2004; Wilson, Jackson & Nichols, 2003).

In order to apply this method, the research team has to choose which of the decision points that have been recorded in observation protocols to analyze in more detail. It is usually not feasible to analyze all decisions. The choice is made based on the relevance of the operational uncertainties that are addressed. Therefore the researchers need to map the decisions to the work domain an control tasks from the previous steps.

The chosen decisions are then retrospectively discussed in off-line semi-structured interviews. The participants are asked to recall their decision and the processes involved. Questions are for example:
- What information do you need to take this decision?
- Was the information readily available? Where?
- Did you contact any other person to make this decision?
- Were there alternative options to choose from?
- How did you choose between alternatives?
- Do you decide differently in other circumstances?
- Where do you have to communicate your decision?

The results of these 'decision probe interviews' are process-oriented protocols of decision-making episodes on the workplace level. They can be related to a general system state (e.g. overload) or typical ways of functioning within the organization (e.g. 'last minute orders').

6.1.3.6 Step 6: Interdependencies of control tasks (information flow, feedback)

Decisions made in specific situations usually become the background for future decisions. Decision making is the core function of organizations (Luhmann, 2000). Therefore, an information flow between control tasks is necessary and thus can be targeted in analysis. Control tasks are forming a network that generally also serves distributed decision making.

A description of the inter-connectedness of control tasks and activities allows to reveal 'typical' modes of functioning in control, e.g. what some planners call 'fire-fighting'. These modes of control vary depending on the workload and other environmental factors leading to time constraints and other limitations in the general functioning of the control system (cf. Hollnagel & Woods, 2005, p. 157ff).

6.1.3.7 Step 7: Used vs. potential control possibilities and capabilities (competencies)

Coming from the analysis of single control tasks and decisions, a network of control functions can be described that acts in a more or less adapted way to meet the control need or demand imposed by operational uncertainties. The control system (i.e. the network of control tasks and responsible actors) therefore exploits the control possibilities and capabilities available. The sum of these are forming the general competency of the system to control uncertainties.

However, there might be control possibilities and capabilities that are not exploited. These are addressed in the final step of the analysis through interviews on the management level. Questions are concerning the reasons and limitations in the exploitation of control possibilities (i.e. technical) and capabilities (i.e. qualifications). This step is resulting in a final evaluation of the used and unused control opportunities including a list of potentially available control possibilities as well as capabilities that have a potential for development within the company.

6.2 Data collection

Three different types of organizational stakeholders were involved: Planning, engineering and production. All three departments provided a key informant for our study. A total of 10 semi-structured interviews and 3 systematic observations on site were conducted. The data was recorded and transcribed, which resulted in 131 pages of raw material for our analysis. Table 18 is showing an overview of our data collection activities.

Table 17: Interviews and observations in case study two

Nr.	Department	Interview duration	Topic	Observation at workplace
11	Planning	1:39	Work domain	-
12	Production	0:58	Work domain	-
13	Engineering	1:02	Work domain	-
14	Planning	2:58	Operational uncertainties	-
15	Engineering	2:00	Operational uncertainties	-
16	Planning	2:38	Control tasks	-
17	Engineering	1:59	Control tasks	-
18	Production	0:40	Decisions	3:00
19	Planning	0:11	Decisions	2:45
20	Engineering	0:56	Decisions	3:00

Study two 165

The following chapter will summarize the results of the seven analytical steps that were employed in our second case study. This case will then provide the material for a subsequent analysis and theoretical discussion within the perimeter of this thesis.

6.3   Results of (extended) Cognitive Work Analysis

The company, which is embedded in a large international corporation, started 10 years prior to the study with a staff of 30 persons, at the time of the study they are operating with 150 employees and a yearly turnover of 50 million Swiss francs. Their service offers consist in the refurbishment and reconditioning of rotating as well as casing parts of gas turbines that are used to produce electricity. The center of operations is in Switzerland, with several accessory workshops abroad, to save costs. Some orders are split between the various sites, which creates additional complexities for planning and scheduling. Planning tasks are concentrated in Switzerland, scheduling is done individually at the different sites, but not for all tasks. Between 40'000 and 50'000 parts are reconditioned every year.

The process is divided in five sub-processes. First comes (1) a pre-assessment of the condition of the parts. After (2) cleaning, a detailed assessment follows, and the (3) repair process is triggered. After various steps of repair and refurbishment, the parts are (4) coated again and (7) quality is assured before shipping. 40-50% of value creation is in-house, the rest is done through suppliers, for example strip-cleaning or coating. The reconditioning of casing parts is done in a separate workshop since it usually consists of highly customized work that needs specialized engineering and craftsmanship.

At the time of the study, delivery reliability was 90%. On average, delayed parts are shipped with 11 days of delay. In earlier years, the delays were much more of a problem. Three years prior to the study, 1.5 new posts were created for production planning. Prior to that, there was only a 0.5 post allocated for planning. Processing time is 4 to 6 months. Volumes have increased substantially during the past years. The company can handle about 60 new orders per month with average difficulty in terms of reconditioning. The amount of work varies enormously depending on the general condition and degree of damage of the parts that are delivered.

The degree of damage is related to the modes of usage by the customer. If a turbine has been running on and on for some time, the parts are in a much worse shape that if the turbine has been used to support other power plants in times of increased electricity demand. Although the gas turbines in use are mostly known by the producer, who keeps records of all installations, forecasting is difficult. Also, there are competing companies that also offer reconditioning. Some customers have operational maintenance contracts. Market share is between 20-100%, depending on the type of turbine. It is generally lower with older, simpler and smaller turbines.

According to the CEO of the company, the challenges remain, even with the new enterprise resource planning system fully implemented. Constant adaptations are necessary, only human cooperation and experience ensure timely delivery. The dependency on the system on one side ("everything is in there, with best data quality possible [personal notes, translation by author]") and on human collaboration on the other side ("not true, come see on the shop floor [personal notes]") creates potential for conflict. Both ways of thinking have their strengths and weaknesses. A common and integrating approach to create a well-accepted socio-technical system is desirable according to the CEO.

6.3.1 Work domain

A workshop with key persons revealed five core processes in planning and scheduling:

1. Intake of orders: Based on the pre-assessment of the parts a preliminary production schedule is established. A production order is issued and a delivery date is set.
2. Planning of repair works: Based on the detailed assessment of the stripped (cleaned) parts, the production order is changed. The number and type of damage is heavily influencing repair time and scheduling.
3. Re-planning of repair works: Due to non-conformance reports, repair works have to be rescheduled. This can be due to irreparable parts that need to be scrapped or due to a more complicated repair effort that is needed.
4. Capacity planning: Every workshop within the factory (welding, grinding, sanding etc.) has its own capacity planning for a time frame of about four weeks.

5. Daily scheduling: Based on the tasks, non-conformance reports, disruptions of production and skill-grade mix of the workers present during a shift, the foreman has to adapt the work plan on a daily basis.

Production planning and scheduling is established collaboratively mainly through the planners and schedulers as well as the foremen, with the support of the engineering department. There are more than 30 external suppliers of parts and services involved. Customer care is done through the sales personnel, which is not within the company. Sales is allocated on the corporation level. Customers are sometimes allowed to visit 'their parts' in the factory during the reconditioning process.

6.3.2 Operational uncertainties

There are many operational uncertainties involved in the production process. First, the exact type and scope of damage is only revealed after stripping and cleaning of the parts. Some problems occur only during the repair process, e.g. if venting holes are closed due to welding and need to be redone later. Some parts are at the beginning believed to be reparable, but then need to be scrapped later on. Suppliers deliver not only raw materials (e.g. for welding) but are also service partners (e.g. machinery) and members of the supply chain (e.g. chemical stripping, coating). Quality assurance is a separate department within the company that takes care of quality questions and milestone testing during the production process. The uncertainties of the repair process itself as well as the multitude of supporting processes and partners creates a very high complexity for planning.

When the enterprise resource planning system first was implemented, huge efforts were undertaken to map the actual production process. A standard procedure (called 'normal work plan') was established where durations for every single step of the production process were recorded. This approach was subsequently modified and changed. One of the fundamental problems were the deviations from mean production time, especially in (manual) welding and grinding that require a lot of skilled labor. Therefore many production intervals were furnished with 'dead times' to ensure sufficient 'slack' in the standard procedure. In doing so, the frequency of re-planning was reduced.

Today, there is a large decision latitude on the shop floor, with no binding job sequence any more. The foreman receives a list from the system with jobs that need to be done or started sometime soon. He then choses from the list what seems feasible to him given the actual circumstances on the shop floor. Every production procedure has to be 'recorded' using a bar code on the order papers. However, this is often done to satisfy the requirement, not to really start working on the job. Also, delays are often not reported, which creates planning problems later.

Twice a week, production planners are meeting with purchasing agents. Once a week they are meeting the production manager as well as the foremen from the workshop. These meetings are held to ensure communication about delays and anticipated or already existing problems. There are certain bottlenecks in the production process that need to be proactively dealt with, for example the oven that is used for hardening. A detailed compilation of all aspects of the planning and scheduling work domain (decomposition-abstraction matrix) can be found in the annex of this thesis (cf. Table 40).

Since the corporation develops new and ever more elaborate kinds of gas turbines for power generation, there are also new kinds of parts to be reconditioned. The problem is that there is not much knowledge about feasibility and efficiency of the processes needed. The engineering department therefore is responsible for stop-and-go decision making in collaboration with production and quality assurance teams. This complication is a very big challenge for production planning and scheduling, since there are frequent stops to do tests or calculations for risk assessment. Reporting and documentation requires a lot of time as well.

According to the shop floor supervisor, there are also problems due to "urgent orders [12]" and the splitting of batches to speed things up. This leads to logistics that are not 'standard procedures' with the risk of mistakes and chaotic local management. I have even seen cardboard boxes with parts under the table of the supervisor. One more difficulty is the lack of visibility of most parts, since they are stacked in special wood boxes that need to be opened to inspect every part individually in case of identification problems.

The task of the production planner was qualitatively different at times. Depending on whether the factory was running at full capacity or having too little work load, the planner needed to apply different strategies to achieve his goals. In the first case, he was constantly correcting and adapting the plan due to uncertainties in the production process. He was very actively planning the bottlenecks. In the second case, he was busy 'shortening' production by removing slack in the 'standard repair procedure'. In this way, he was able to artificially raise workload locally, e.g. in the welding shop.

Some of the problems that the planners had to deal with were related due to a time delay in the actualization of the system. This lead to the phenomenon that jobs that were reported 'done' in one shop did not appear on the job list of the subsequent shop as 'ready'. The planner then had to personally inform the shop floor supervisor or the foreman that his men can start to work on this job. In normal circumstances he would not do this but it became an important time gain in hectic situations.

In general, the feedback system was designed in a particular way. It would only allow to report 'used time', not the actual fulfillment of the job. This regularly led to the situation that the time for the job had elapsed, but the job was not finished in reality. It therefore took careful consideration to interpret this date. It was possible that the workers did put much more effort in one order, but this was not visible in the system. There would just be a 'zero' for 'time to job fulfillment', and no 'negative' time even though the workers would continue to work on the job order. The planners had to inquire with the foreman or the shop floor supervisor to get an accurate picture of the actual situation. Very often, the shop floor supervisor would not proactively report such unfinished or overdue jobs.

The dispatching of workers to specific jobs was done through the shop floor supervisor. It was not documented at all, since he did so 'mentally' and without any kind of tool. It seemed to be very common that on the work plan produced by the system were four or five jobs that need to be worked on at a particular day, but in reality all the workers were allocated to one single job.

The workers register every work process with the computer using a bar code. It is allowed to register work on a job order that is still 'active' in a previous production process or shop. This is necessary for example when parts have to be welded and grinded over and over again to achieve the necessary standard in quality. However, this practice is leading to a lower transparency as to where exactly the work progress is at a given moment in time. The planners have to actively inquire about a specific job order. The shop floor supervisor has a certain pressure to make sure that every shift has enough work 'ready' and parts as well as raw material in place. This is achieved through a structure in the time recording system – whenever a worker is 'idle' the recorded time is booked on the 'account' of the shop floor supervisor. Therefore, some job orders might already be 'started' in the system, but in reality are only serving as "reserve [12]" to buffer for unwanted "management attention [12]" towards the shop floor manager.

All the workers in the production were subject to a collective labor agreement, working in two shifts. There were defined annual working hours allowing for a certain flexibility on the side of the company. The planner was allowed to increase capacities if necessary within certain boundaries. In this way, working extra hours on Saturdays and even Sundays was possible. These extra hours were very attractive for workers since they paid higher. The relatively low wage motivated many workers to willingly put in extra hours. It was suspected by management that work sometimes 'took longer' so that it became necessary to finish on Saturday morning. When the new manager had decided to scrap the supplement payments for Saturday and Sunday work hours (without raising the average wage), he had faced stiff resistance on the side of the workers. He had had to renounce from his plans. This meant that the salary structure of the company had direct consequences on production planning.

Since for this case study, an evaluation of the control capacity was our main objective, a separate chapter will summarize and discuss the findings regarding this conceptual model. To ensure readability, I decided not to include details of the findings here in this chapter. The interested reader will find some examples in the annex, however in German language.

Table 18: Resulting control capacity overview

	System level	**Workplace level**
Control needs	Coping with high uncertainty regarding the actual amount of repair work needed for a particular job; Large number of dependencies in the supply chain; Circular processes on shop floor (production and quality control).	Allow for a big enough choice of job orders to account for available qualifications in the work force present in the shift; Avoid management attention due to 'worker idle time'; Monitor delayed orders in production.
Control possibilities	Enterprise resource planning allows for control of interfaces (sales, purchasing, delivery); Standard work plan that is adaptable when knowledge is available (due to assessments or testing).	Daily job lists with time frame of several weeks; Work activity registration allowed for multiple shops simultaneously; Splitting of orders; Overtime on Saturdays and Sundays; Modification of standard work plans.
Control capabilities	Trained planners; Use of planning module to schedule job orders; Control of job routing / sequencing depending on general workload.	Massive allocation of resources to single jobs on the shop floor; Tracking and 'chasing' of individual job orders; Off-line optimization from one shift to the other.
Unused control opportunities	Structured, transparent job order-related communication of operational uncertainties; Worker participation in structural decisions on control.	Visibility of work progress; Distribution of job allocation decisions to enhance transparency on shop floor.

6.3.3 Control capacity

As we have laid out elsewhere in more detail (Karltun et al., 2006; Wäfler et al., 2008), the aim of our analysis was to measure and evaluate the potential for control or control capacity of a company using a structured empirical approach based on the framework of Cognitive Work Analysis – with some extensions.

The final result of our analysis was an overview of the control needs, the control possibilities and capabilities that were in place on the system as well as on the workplace levels of the company. Table 19 is showing these findings, including an evaluation of unused control opportunities, according to the research team.

6.4 Morphogenetic analysis

The fundamental question that has not been addressed in the first analysis of the material that had been collected in this case study is the question about *why* some control opportunities are exploited and others are not. Whilst it is certainly possible to further explore the *constitutive entanglement* of the social and the material, this approach will not reveal the underlying design principles or restrictions (Mutch, 2013). In order to do this, I will have to take another route, focussing on the characteristics and features of the control *regime* that is put in place using various technological, psychological and organizational means. This structure has a history, and it has emergent properties of its own.

6.4.1 Questions to ask

The morphogenetic approach requires that five *basic questions* are answered regarding the structure(s) of interest and its (their) causal power(s) (Elder-Vass, 2007, p. 31):

1. What are its characteristic parts?
2. How do they have to be structured (i.e. related to each other) to form the whole?
3. How does this come about (morphogenesis)?
4. How is it sustained (morphostasis)?
5. How are the powers or properties of the whole produced as a result of it having the parts it does, organised as they are (explanation of the generative mechanisms)?

Study two 173

In the following, I am attempting to analyze our empirical material according to these questions. To structure my text I chose to use the temporal order proposed in the original approach by Archer (1995), shown in Figure 10 further above.

6.4.2 Structural conditioning / inscription

Volkoff and her colleagues (2007) have been able to show that the characteristic parts of ERP systems are *embedded organizational elements*, such as routines, roles and data as well as their relationships [cf. basic question number 1 by Elder-Vass, 2007, p. 31]. Relationships among routines form work flows or sequences, roles are based on their relationships to transactions (i.e. routines) and there are relationships between transactions and data (i.e. data requirements to execute a certain transaction). However, an organization is not completely free in implementing or embedding its routines and roles. As Volkoff, Strong and Elmes put it: "The routines enacted through social interaction during ES [Enterprise System] use do not construct the system – the system exists prior to its use, having been designed and built by a vendor, and configured by an implementation team. These design and construction and configuration activities create the prior structural conditions within which use eventually occurs (2007, p. 844)."

As Leonardi defines, "sociomateriality is the enactment of a particular set of activities that meld materiality with institutions, norms, discourses, and all other phenomena we typically define as 'social' (2012, p. 24)". In critical realist terms, sociomateriality employs a 'flat' ontology, because it is conflating agency and structure. The crucial question, however, is the one about *temporal* order: Who has established the 'set of activities' in the first place? Who decides what can be changed in a 'sociomaterial' system, and what not? Where do norms and discourses come from? Or, more in general, what are the preconditions and requirements that need to be met for sociomateriality to actually take place [cf. basic question number 2]?

Thus, along the lines sketched out by Blumer (1990, p. 27f) in his very interesting discussion about the notion of *industrialization*, we could inquire more critically about the introduction of Enterprise Resource Planning in manufacturing and its conditioning effects [cf. basic question number 3]:

1. Where has the ERP been developed, by whom, for whom and what are the consequences of this regarding the form and functionality of the system?
2. How much tailoring is possible, what does it cost and who decides on these expenses? Is new hardware required?
3. Who is training and motivating people that are supposed to work with the system?
4. Does the ERP system work without any other adaptations, e.g. on the side of suppliers, subcontractors or transport partners?
5. Is a company free in choosing the service partner or bound to use certified personnel by certain providers of services and support?

The decision to implement an expensive and all-including software system such as an ERP is not taken lightly and usually involving the top management if not the shareholders or owners of a company. In any case, it requires high-level structural powers that allow for the subsequent change of rules and standards related to the most important processes in the organization. This – at least potentially – might also create resistance and conflicts, since "the power to do something (…) is always the power to do something against someone, or against the values and interests of this 'someone' that are enshrined in the apparatuses that rule and organize social life (Castells, 2009, p. 13)." Another question to ask therefore would be, who is profiting and who is loosing in ERP implementations?

Origins. The term ERP was coined by the Gartner Group, a consultancy firm, in the early 1990s. Another name, CAPM (Computer Assisted Production Management) was in use at the time, but eventually was dismissed (cf. Webster, 1991a). Gartner's consultants described a software architecture – which they called ERP – that they considered the next generation of MRP-II in April 1990 (Pollock & Williams, 2009, p. 24). The architecture is based on a vision of enterprise integration with a three-level client-server architecture (Pollock & Williams, 2009, p. 25). Today's ERP systems are still based on the MRP-II framework (cf. Figure 5 above). In our case, the corporation had decided to use one of the most common large-scale ERP systems on the market. We did not inquire into this specific topic but can assume that the most important aspect was integration of all divisions of the corporation.

Segment. Advertisements, supplier literature, demonstrations on trade fairs, reports in trade papers all play a role in the acquisition process. Once the vendor and the customer have decided to work together, a tight contractual relationship is formed. There might be suppliers of associated products involved, i.e. hardware. There often are external knowledge providers that offer training and consultancy services. There might be implementers, suppliers of complementary or interoperable products, system integrators, change management experts etc. involved as well. The customers might participate in user clubs, usually based on contacts within a certain industry (Clausen & Koch, 1999; Koch, 2005; Pollock & Williams, 2009). Clausen and Koch (1999) have introduced the term 'segment' to designate such conglomerates of IT suppliers and customers. They suggest that knowledge flows within these segments – and is forming the evolution of ERP systems through implementation experiences, new demands and visions. In our case, the decisions concerning the ERP system were taken within the corporation as a whole, not the individual company. Support was available from the headquarters, but only on a limited scale. Our interviews did not cover these aspects of the IT infrastructure, but we can assume that most if not all of the above partners and participants were involved at some point in time.

Features. The functions and features reach from analytics (financials, human capital, procurement, product development and manufacturing, sales and services) to strategic enterprise management (financial supply, supply chain management, financial accounting, talent management, workforce process management, inventory and warehouse management, production planning and manufacturing execution, sales order management, aftermarket sales and management, real estate management, enterprise asset management, travel management, environment) and to operations and workforce analytics (management accounting, corporate governance, workforce deployment, logistics, transportation management, product development, lifecycle data management, health and safety, quality management). In addition to that, providers are offering industry-specific packages and 'business maps' (e.g. for financial services and banking, manufacturing or service industries, hospitals and higher education).

Historical context. Four main factors contributed to the expansion of ERP use: (1) technical development, i.e. the availability of faster, cheaper computing capacity and shortage of IT professionals, (2) managerial, i.e. BPR and lean manufacturing ideas, 'best practice', (3) desire to share information across the whole company in combination with more financial control, i.e. an integrated system approach, (4) marketing, fear of Y2K problems in older systems. In addition to that, "ERP offerings with their libraries of 'best practice' industry business processes could be presented as a vehicle for BPR, mainly by consultants and vendors (Pollock & Williams, 2009, p. 28f)".

Acquisition. Procurement of ERP systems involves top managers and IT specialists, and the process is generally lengthy and usually not straight forward. There is a considerable amount of 'drama' involved. Not only the status and provenance of the software, but also the credibility of the vendor, the judgments of experts, the availability of successful cases of the same industry as well as convincing demonstrations are contributing to the decision-making process (Pollock & Williams, 2009, pp. 60f and 176ff). None of these factors or processes is known in our case, since we got in contact with the company some time after implementation.

Change. Any particular moment in the history of an organization can bee seen as simultaneously involving the unfolding of: (a) a task, (b) an individual project, (c) an occupational project, (d) an organizational project, and (d) the production or reproduction of an institution (Pollock & Williams, 2009, p. 279). All these simultaneous processes that might or might not unfold at the moment of the implementation of a new technology are affected in some way or the other. ERP systems support certain ways of organizing and inhibit others (cf. Koch, 2005; Webster, 1991a). In our case, there were substantial changes on the level of the shop floor when the system was first implemented. The shop floor supervisor lost almost all of his decision latitude. He now received detailed daily work plans, generated by the algorithms of the ERP system, that had to be executed without further delay. There was very limited flexibility in terms of allocating qualifications and personnel present in a specific shift to specific order demands.

6.4.3 Social interaction / agential reflexivity

Other research teams using the case study approach have been able to show that already during implementation the participants were forced to 'work around the plan' to obtain solutions to emerging issues. In using a different but comparable example, Elbanna demonstrates that improvisation, bricolage and drift took place on both the technology and organizational side at the company they studied (Elbanna, 2006). These findings indicate that the negotiations between organizational actors begin at a very early stage of interaction with the system [cf. basic question number 3]. As a matter of fact, through the inscription and subsequent empowered or impaired modes of interaction, the various interests within an organization are mobilized. In other words, as Castells argued "... power is not located in one particular social sphere or institution, but it is distributed throughout the entire realm of human action. Yet, there are concentrated expressions of power relationships in certain social forms that condition and frame the practice of power in society at large by enforcing domination. Power is relational, domination is institutional (Castells, 2009, p. 15)."

When it comes to social interaction in and around ERP systems, power and domination are important underlying processes. In our case study we were primarily interested in the thoughts and concepts about the workings and failures of the planning system held by the participants. In order to elicit these understandings, we conducted three subsequent workshops following the method that initially was described by Vester (2005). 20 variables were discussed and described by the interdisciplinary participants of these workshops These factors or variables are thought to have causal effects on the degree of coordination of job order dispatching and control (a detailed list including corresponding indicators can be found in the annex). They included for example incorrect planning parameters, degree of collaboration in production (among shops), degree of collaboration between production and engineering, proximity (physical, organizational), worker qualification, work execution, independence from key persons, common understanding of process definitions, capability and capacity to execute urgent orders, technical and quality problems, undue management interventions, and others more.

There are three categories of powers at work that shape – but not determine – social events: Social structures, individuals and other material objects such as technological artifacts (cf. Elder-Vass, 2010). Table 20 is showing the findings of our case study according to these three sources of powers.

Table 19: Variables influencing coordination of job orders

Powers of social structures	Collaboration in production (among shops), collaboration between production and engineering, proximity (organizational), common understanding of process definitions, independence from key persons, capability and capacity to execute urgent orders
Powers of individuals	Qualification, work execution, management interventions
Powers of other material objects	Planning parameters, technical and quality problems, proximity (physical)

Through the lens of organizational routines (Feldman & Pentland, 2003), some of these elements or influences clearly have 'ostensive', and others predominantly 'performative' aspects. Common understanding and capability to execute urgent orders are based on ostensive knowledge within the workforce of the company. As opposed to that, proximity and management interventions are performative in nature.

In our case, the consensual identification of influencing factors and variables had immediate consequences within the company. The management decided to divide the engineering department into a development group and a production support group. The production support engineers were then relocated into a newly created office space right above the shop floor (not in the office building next to the factory). This allowed for proximity and a much more coordinated collaboration between the production and engineering departments. The aim of these measures was to reduce answer latency when questions arose during the assessment or repair process. Most importantly, the higher availability of engineers was intended to help speeding up decision processes in regard to new products that had never been repaired and reconditioned before. There was some resistance from engineers, but eventually they agreed to the change in their responsibilities and duties. No major change in the ERP system was needed to establish this new production support team. However, the overall ERP functioning possibly improved through the production

Study two 179

engineer's correction of missing or incorrect parameters within the production plan.

One of the identified problems persisted over some time. Due to a very limiting daily job list, the shop floor supervisor was frequently not able to allocate the sufficiently qualified man to a job, something which is especially important in welding. Welders have very different levels of skill and expertise and some are qualified to do specific repair work whereas others are not. With a daily job list but shifting personnel, the supervisor and his foremen were put into a very difficult position. This arrangement would only work with a homogenous team of welders, for instance. This is not feasible at all, since there is a shortage of experienced welders on the market.

The erosion of competencies on the level of the shop floor supervisor (in German 'Meister') has been described by Kleinau (2005). He found that the role of the shop floor supervisor has been undergoing substantial changes through four main influencing factors: (1) Lean management and process-based responsibilities, (2) further automation and introduction of planning algorithms, (3) further division of labor, e.g. quality assurance or maintenance, and (4) introduction of group work and thus less hierarchy on the shop floor.

In our case, the ERP implementation was following a rather 'weak' concept of team work, with very limited autonomy and decision latitude on the shop floor level (Koch & Buhl, 2001). This certainly is the management approach that is requiring the least amount of adaptions within the ERP system [cf. basic question number 4].

6.4.4 Structural / sociocultural elaboration

The problems related to the dispatching of jobs to sufficiently qualified personnel through the shop floor supervisor was eventually solved through the extension of the 'normal work plan' and the introduction of a much greater flexibility of work plans. Essentially, the work plans were extended to a weekly timeframe, although they could still be actualized on a daily basis. The enhanced decision latitude on the shop floor corresponded to the requirements for the management of critical repair processes but it also decreased transparency on the side of the planner. In the newly defined mode of job allocation, it was much more difficult for the planners to actually establish a sufficiently detailed overview of the daily 'state of affairs' on the shop floor.

However, the higher degree of autonomy on the shop floor level was not awarded to empowered teams ('strong' mode of team work, cf. Koch & Buhl, 2001), but remained with the shop floor supervisor, who would allocate jobs to persons 'in his head'. In this way, the work-around found by the planners and the production manager was increasing the communication requirements between the planners and the shop floor as well as increasing the dependence on the shop floor supervisor himself [cf. basic question number 5].

As Martín-Baró has aptly pointed out, the domain of work, including services and industrial production is *per se* political. He wrote: "Few would deny that the underlying power structure of a society comes into being through economic relations that dictate the social division of labor. If, then, power is rooted in the resources available to groups and people as they carry on in the diverse realms of social life, it is obvious that the world of work constitutes the political sphere par excellence (...). As people act to get their human needs met, social lasses take shape, through mutual interrelation with the boundaries defined by the modes of production. In this same process, individuals flourish in their development, or they stagnate; they achieve their hopes, or retreat into alienation: they are humanized, or dehumanized (Martín-Baró, 1994, p. 97)."

In the company we studied, there was a sort of 'cultural divide' between the 'white collar' workers and the 'blue collar' workers. Managers, planners, engineers were predominantly Swiss or German. On the shop floor the majority was from other – mostly southern – European countries as well as Turkey. The shop floor supervisor (himself from Germany) mentioned that they were from 15 different nations, and that there had only been one violent conflict so far (!). Clearly, with this divide in place, there was not much resistance nor demand for participation in the structuring of the company. In other contexts, with a higher degree of union memberships and less cultural differences, there might also be more critical voices. These interest groups could demand more autonomy and a more 'human' mode of collaboration, but this was not the case, at least not during the time of our inquiry at the company (cf. Koch & Buhl, 2001).

Implementation studies that end too soon may underestimate the eventual organizational consequences of an innovation (Pollock & Williams, 2009, p. 85). The same could be said of our own study. Following our report to the top management, we were invited to

Study two 181

participate in a meeting concerning a new add-on for the ERP system. This add-on was promoted by the CIO of the mother corporation, who was present at the meeting. A sales representative of the vendor of the tool was presenting the features and capabilities of the software. The CEO of the company wanted to have our opinion on the tool, and whether we thought it was of any use for the solution of their own problems in planning and scheduling. We made some critical remarks and questioned the tool, basically because of its 'data-driven' approach (which requires exact data). Although everybody at the company knew that the usefulness of the add-on was very limited, the tool was purchased and installed.

What became obvious in these discussions on ERP related issues with production planning and scheduling was that the influences were not only within the company itself. As Koch states, „in parallel (...) user groups, consultancies and others develop interpretations and political programs on what is now needed (2005, p. 53)." These networks, or segments, may come to the conclusion that more of the same is needed. More data, more algorithms, more graphical user interfaces. In a similar way, ERP systems get extended with internet portals, supply chain tools, and more (Pollock & Williams, 2009, p. 125f).

There are three possible strategies for influencing the further development of mass-customized ERP systems 'from below': (1) create success stories and exchange templates of ERP-configurations that support users, (2) assure continual improvement of ERP-configuration, (3) create a multinational actor that can influence the generic part of the system, especially also in new sectors that need to be developed (cf. Koch, 2005; Locke & Lowe, 2007).

6.5 Intermediate discussion

The data collection for study two was conducted using a 'modernist' perspective. Its objective was to contribute to the solution of a problem that is underspecified and allocated in a highly 'intentional' work domain, namely production control of discrete manufacturing. The related – mostly cognitive – work that is performed by the organizational actors involved, i.e. the execution of control tasks to cope with operational uncertainties, is partially invisible and disembedded (cf. Star & Strauss, 1999). Such work is subject to constant negotiations between the demands and requirements of different interest groups as well as the control demands of the actual work system. A purely modernist or normative-descriptive approach (cf. Vicente, 1999, p. 61ff) will struggle to produce a sufficiently adequate representation of such a work domain with all its human, social, cultural and material subtexts.

In terms of methodology, it was therefore not feasible to follow the original CWA approach that was developed mainly for 'functional' work domains, albeit from a 'formative' perspective which we considered suitable *in principle*. To start with, the formal description of the work domain provided a challenge in the case of the intentional and widely distributed, hybrid domain of planning and scheduling in manufacturing. Even more so in this particular company with very complex processes (job shop with very high operational uncertainty due to repair work with unclear scope). The initial analysis of our data therefore had to remain partial and was not fully satisfactory. It provided some insights in the ongoing process of organizing, but did not help to clarify other aspects, like negotiations *about* control within the company. For the purpose of this thesis, however, I believe the study provides invaluable material. It allows for a reconsideration of the fundamental theoretical assumptions connected to such 'application-oriented' psychological studies of work contexts that are conducted with an intention to develop or improve technologies such as decision support or virtual team collaboration systems.

The focus of the *subsequent* analysis that I conducted for this thesis was on additional explanatory capabilities (cf. discussion of study one). Based on a realist ontology, the morphogenetic approach allows for the description and understanding of power-relations related to technology development, implementation and use (Elder-Vass, 2012b; Mutch, 2010). Indeed, as students of work and industrial production, we do have to account for the role of power as a major context influence at play. Not doing so would mean that we do not fully understand human agency and behavior in this kind of context. Our findings would be very limited, or restricted to areas where humans are consciously subscribing to a certain power regime, such as in military contexts or in other high-risk, high-stakes environments (e.g. aviation or nuclear industry).

Therefore I suggest that a morphogenetic analysis of the major structural elements of a work system, such as a ERP system, is an undertaking that is paying off. In our example, case study two, the mechanisms generating opportunities and threats for socio-technical design became more visible and understandable.

As I have attempted to illuminate above, 'packaged' ERP systems are emerging from large and widely distributed networks of producers, vendors and consultants. These networks can be conceived as forming 'sectors', especially for specific countries, industries and/or (work) cultures. The 'packaged' software can evolve to a certain extent. Whereas the producer has interest in offering a standardized and customizable tool, the vendors and consultants (as 'value-adding resellers') are interested in providing guidance and know-how in terms of tailoring the standardized product. Thus, they are not necessarily interested in too many modules and features offered by the producer but instead are offering their own industry-specific solutions as add-ons to the standard software. The customer as well as the end users working for that company generally have little influence on the product itself. They are highly dependent on the other players of the network to provide upgrades, solutions and new features.

The standard software allows for some kinds of organizing and collaboration whereas it hinders or inhibits others. In our case, the core concepts of the company regarding organization were in line with the ERP that had been chosen. Besides a hierarchical organization with distinct departments, it was using a rather 'weak' approach to team work, with little autonomy on the group level. Other organizations might face difficulties or substantial problems in the roll-out process if there is no 'fit' between the general approach to organizing and the ERP product. Under certain circumstances, one of the solution providers that specialize in add-ons might help to establish the necessary modifications during implementation.

There remain many questions as to how the structural elaboration is taking place after 'going live'. It might be interesting to look at various ways of feedback generation, participation, community building and 'clusters', 'segments' or network alliances within an industry.

ERP projects often stretch across long time-spans, from the initial idea to the final steps of the implementation, sometimes 10 years or more, and frequently with a kind of trial and error approach (e.g. Wagner et al., 2010). In our case study one, the company had chosen an ERP system under considerable time pressure because of the Y2K problem. They then realized that this product was not well supported and did not satisfy the needs of the company in terms of flexibility and adaptability. The long time frames and dynamic relationship between different actors pose some problems to apply the morphogenetic approach. For most research teams it is not possible to conduct research in multiple places and across many years. This would demand a dedicated research program with subsequent projects and sufficient funding.

Even if such a research endeavor could be envisioned, not all data might be available or accessible that a thorough morphogenetic approach requires. Especially value-adding resellers are in a position where their product- and industry-specific knowledge is their competitive advantage. They might not be ready to contribute to such an undertaking. The same of course is true for many companies in the industry – ERP system implementation and production planning and scheduling is affecting them in their organizational core. It is not likely that they allow for outsiders to take part in every decision process. The time span, the distributedness of the processes involved and the general accessibility to the sites that are crucial for morphogenesis of such systems are limiting factors for research in this field.

Despite these difficulties, my project of developing a critical theory/methods package for Human Factors, I believe, is a step closer to the intended outcome. Using this case study, I have been able to show that the CWA methodology can be extended to be used in 'intentional' work domains as well. The fundamental ontological position, Critical Realism, is in line with both CWA as well as the morphogenetic approach. In combining the two, I have been able to identify emergent properties, powers and generating mechanisms that help to explain social events and individual behavior in the production planning and scheduling work domain that is full of 'soft factors' and conditioning contexts that reach far beyond the immediate work environment.

In the final chapters of this thesis, I will review the results of the case studies in the light of the intended general framework or theory/methods package. And finally I will discuss practical consequences of these findings for Human Factors.

7 Overall discussion and conclusions

> Insofar as we continue to see the technical and the social as separate domains, important aspects of these dimensions of our existence will remain beyond our reach as a democratic society. The fate of democracy is therefore bound up with our understanding of technology (Feenberg, 1999, p. vii).

What can be said about the development of psychological research in production planning and scheduling after these two exemplary case studies? Do we know more about the shortcomings of a 'modernist' approach in describing and analyzing ERP systems or similar socio-technical assemblages? What are – in the light of my findings – the most crucial characteristics of this particular constellation of 'distributed action', seen from a critical-emancipatory angle? These and more questions will be addressed in this chapter.

In their overview of Engineering Psychology and Human Performance, Wickens and Hollands (2000) refer to social structure only in two aspects, without any further elaborations. The first passage is concerned with communication issues in teams with heterogenous authority ('crew resource management') and the second with errors in organizational contexts ('resident pathogens' or latent conditions for errors). In another reference book edited by Hendrick and Kleiner, titled *Macroergonomics* (2002), social structures are mentioned in terms of professionalism, value systems, cultural diversity, gender, cognitive complexity, participation and organizational culture. However, there is a lack of theoretical embedding of these concepts. They are mentioned as influences, but it remains very vague through which mechanisms they affect individual behavior at the workplace. In a more recent German overview on Human Factors, there is substantially more emphasis on group dynamics and cultural aspects of safety (Badke-Schaub et al., 2008). But here, too, the enormous challenges of distributed and complex socio-technical systems is mentioned only briefly in a chapter on new forms of collaboration (Lauche, 2008). I suspect that one of the underlying reasons for the rather reluctant way of handling issues of social influence in our discipline is that Psychology in general does not consider social structures as causal powers in their own right, but merely as 'independent variables'.

7.1 Social structure in Human Factors

One could conclude that the of consideration of social structures in mainstream Human Factors is not seen as having the potential for important contributions to the discipline. For many fields of application, this may be the case, especially work domains with highly structured hierarchical organization, clear-cut role definitions and a predominantly 'functional' technological environment (e.g. air traffic control, train scheduling or chemical plant management). However, in work domains with a less structured workforce, flexible and dynamic social relationships and roles as well as a predominantly 'intentional' character of the system, the lack of theoretical fundaments in the area of social structures becomes a deficiency for Human Factors. My overview of HF research in production planning and scheduling in discrete manufacturing has illustrated these shortcomings in a specific field of application.

The challenge, then, was to identify exiting methodologies that would allow for an extension and theoretical reformulation to create a critical theory/methods package that could serve as an research approach for exactly those domains that are not well covered with standard Human Factors methodology. After reviewing various contributions that have employed critical and emancipatory thought in relevant fields of technology studies, and some subsequent philosophical elaborations of the fundamental questions raised by this review, I chose a critical realist perspective to develop further one of the existing approaches to work analysis, namely the Cognitive Work Analysis framework.

In his initial formulation of this framework, Rasmussen did not mention the social aspects of work domains at all (Rasmussen, 1986b). A few years later, though, social organization and cultural aspects appeared prominently in the analytical framework that his group proposed (Rasmussen et al., 1994). They consider "the work organization (...) as a distributed, self-organizing control system". Thus, the aim of the analysis is "to identify mechanisms that shape this organization and govern the allocation of functions to individuals (Rasmussen et al., 1994, p. 93)". Their analytical approach is based

Overall discussion and conclusions

on an open systems perspective on organizations (Rasmussen et al., 1994, p. 94). Accordingly, they understand organization as the relational structure that is necessary to coordinate the work activities of individuals. In their view, these dynamic structures have to be distinguished from the formal organization which reflects the – more stable – allocation of authority and legal responsibility.

In both Rasmussen's as well as Vicente's concepts of social organization and cooperation analysis (cf. Vicente, 1999, p. 249ff), the focus is on function allocation, division of labor and cooperation. Vicente is putting more emphasis on the emergent and self-organizing nature of systems of collaboration, providing examples and some references to literature. As Vicente states "… there can be little doubt that complex sociotechnical systems are largely open and thereby require distributed, adaptive organizational structures". The question, then is: "How do we deliberately design for adaptive self-organization (Vicente, 1999, p. 253)?" – The solution, they propose, is to distinguish between the content and the form of cooperation. The content meaning the division and coordination of work according to control requirements, and the form meaning its social organization. The first aspect seems well covered with CWA, it follows the logic of the work domain and control task analysis. However, the second aspect remains somewhat unclear. Although the authors agree on the point that social organization (e.g. through management style) can have a major influence on the resulting output, their tools for the analysis of this aspect are rather weak, if not non-existing (cf. Vicente, 1999, p. 271).

I used two case studies from the domain of discrete manufacturing to illustrate possible ways of extending CWA with regard to social structure. Socio-cognitive discourse analysis, on one hand, allowed for the description of local rationalities and intentionalities that potentially moderate behavior of individual agents beyond the actual functional division of work and allocation of control. The morphogenetic approach, on the other hand, allows for the identification of conditioning structures that are embedded or inscribed into (informational) artifacts, their influence on social or sociocultural interaction with these systems as well as the subsequent modification or redefinition of those structures.

My propositions fit well with the 'mission' behind CWA. All three approaches are in line with a critical realist philosophy of science.

CWA is based on a systemic understanding of the world, therefore proposing an ecological approach to work analysis, starting with the physical and social environment in which work is taking place. Vicente is explicitly referring to both aspects as "realities (Vicente, 1999, p. 48)". A complex and stratified reality with multiple determination of natural as well as social events requires a „formative approach to work analysis (Vicente, 1999, p. 109ff)". In consequence, CWA is about modeling behavior-shaping constraints. Discourses and other forms of social institutions or structures by definition are such behavior-shaping constraints (Elder-Vass, 2012b). Whilst this was not an objective of this thesis, the re-consideration of some of CWA's underlying theoretical assumptions as well as methodological approaches in the light of a critical realist conception of science could be worthwhile. The main benefit could be that both the 'functional' as well as the 'intentional' aspects of complex sociotechnical systems could be analyzed in a more consistent and truly interdisciplinary way.

The application of the two selected methodological approaches in the field of production planning and scheduling has shown advantages, but also potential pitfalls. My approach has shown that socio-cognitive discourse analysis can be used to describe local rationalities that are at work in the organizational arena, using the pre-implementation stage of an ERP system as an example. In particular, the identified fields of tensions between position-practices and various social and organizational phenomena emerging from them are not yet fully understood and explained. The methodology is stepping short to reveal the full scope of causal powers employed by the different social structures in place. In the second case study, I have been able to show that the CWA methodology can be extended to be used in 'intentional' work domains. In combining CWA with Archer's morphological approach, I have been able to identify a range of emergent properties, powers and generating mechanisms that help to explain social events and individual behavior in the production planning and scheduling work domain.

But still, the question of how individual agents support and undermine ERP systems at the same time remains to be addressed. A complementary, dialectical perspective might be of use in clarifying these issues.

7.2 Agency in Human Factors

What is agency and who can have agency? In literature, a vast amount of contributions to these questions can be found, going back to the ancient Greek philosophers. It is not my intention to reproduce or summarize those manifold strands of philosophical thinking. But, for the purpose of my discussion of agency within Human Factors as a discipline, I have to make some preliminary declarations. As a bystander of this major debate within Philosophy and Sociology, I can only reflect on my own scientific work and chose my own standpoint accordingly.

Psychologists in general do not invest too much time to think and reason about the philosophical assumptions of their discipline (cf. Teo, 2009; Tolman, 1994b). They subscribe to the 'modern' project of psychological research to explain the 'mechanics of the mind'. The object of their scientific work is the behavior of human individuals that are a-historical and de-contextualized. The preferred mode of working is through controlled experiments and observations. In doing so, scientists are following a reductionist, empiricist mode of knowledge production. The underlying mechanistic world view has been criticized by many thinkers during the past three or four decades, especially also because of the persisting reluctance of mainstream psychologists to reflect upon the socio-historical context as well as political-economic consequences of their science (Fox et al., 2009; Gillespie, 1992; Holzkamp, 1985; Kvale, 1992; Parker, 2007; Riegel, 1978; Teo, 1998; Tolman, 1994b; Tolman & Maiers, 1991). A mechanistic world view leaves an individual's subjectivity in a degraded status, as an irrational interference in the development of 'objective' knowledge. The scientific methodology becomes more important than the adequacy of that approach with regard to the subject matter (Teo, 2009). Critique of such an approach to Psychology has not only been raised from the advocates of subjectivity, but also on purely philosophical grounds (e.g. lack of social ontology, disregard of Popper's falsification theory).

When it comes to ontology, mainstream psychologists follow a *positivist* understanding of science. After the turn towards more critical and dialectic thought in the social sciences, that theoretical standpoint has become 'outdated' in many disciplines, but not in Psychology. Some of the other disciplines, or their proponents, have turned to a radical constructivist position, denying any kind of reality outside human thought and language. Between these two poles, positivism on one side and radical (or 'hard') constructivism on the other, there have been many attempts to revise and elaborate their philosophical fundaments. One approach to the revision of fundamental assumptions, without retreating towards radical constructivism, was a shift towards an ecological approach to cognition (Barker, 1968; Bateson, 1972; Gibson, 1979). Another important approach is based on the phenomenological tradition of philosophy, focussing on the structural analysis of experience. According to the proponents of this school, the irreducibility of the first person, or subject, must be acknowledged (cf. Zahavi 2005).

This view consists of an understanding of the self as a center of intentionality that is relating to a world that is independent and to a large extent unknown. Such a conceptualization of humans as subjects has far-reaching consequences on the notion of agency, and thus for Psychology as a discipline. As Tolman puts it: "Seen in this way, reality becomes for each of us a possibility-space within which we experience and act, and intentionality becomes a possibility-relation to our world and ourselves (Tolman, 1994b, p. 33)." As a consequence, reality, for critical psychologists, does not predominantly consist of decontextualized stimuli, but of meanings. These meanings are the structures responsible for human behavior. „This reflects the fact that the dimensions and extent of individual actions (...) are societally determined, but in such a way that individuals can consciously orient themselves towards the situation as a 'possibility', that is, they always have the alternative of acting otherwise or not at all (Holzkamp, 1992, p. 198)."

Within Human Factors, a substantial part of scholarly work is concerned with predicting human performance within its 'natural limitations', i.e. in processes of perception and cognition. It is therefore based on psychological theories of signal detection, vigilance, attention, information processing, memory, decision-making that are applied to work settings involving technological artifacts and studied under controlled conditions (cf. Wickens & Hollands, 2000). It is not difficult to see that these efforts are based on a similar mechanistic empiricist approach. The main tool for Human Factors is modeling, based on variables that are related to each other. The model building that predominates Human Factors has a high validity in describing and predicting the "local relationships between people and technology,

Overall discussion and conclusions

but support no independent criterion for explaining action outside individual (or small group) intentionality (Pettersen et al., 2010, p. 184)." In the domain of production planning and scheduling, specifically, there were efforts to formulate a 'human model scheduler' (Sanderson, 1991). Interestingly, this endeavor has never come to an end and was eventually dismissed.

Interestingly, there are some domains of Human Factors that have elaborated alternative approaches, mostly driven through practical problems. The researchers involved have taken distance from the goal of exact modeling and prediction of human behavior in complex systems. Instead, they employ various methods to establish analytical accounts of work domains that are populated with knowing and empowered individuals. They acknowledge the degrees of freedom that professionals have in their highly complex tasks and strive to uncover regularities and characteristics in their work. While many research programs stay close to the philosophical ground of mainstream Psychology, others have started to explore more ecological, phenomenological or constructivist scientific approaches that employ quite different ontologies and epistemologies.

Pettersen, McDonald and Engen, for example, are arguing "that when viewed in relation to theoretical antecedents the social theorising of safety in socio-technical systems is dominated by cognitive and constructivist based approaches not capable of capturing the relationship between social structures and individual agency (Pettersen et al., 2010, p. 184)." To my knowledge, there is at least one framework in Human Factors that has attempted to bridge the gap between the various perspectives, Cognitive Work Analysis (Rasmussen et al., 1994; Vicente, 1999). It has achieved that through the combination of an ecological understanding of complex sociotechnical work environments and phenomenological descriptions of actual practice in such contexts. The main proponents of that approach have implicitly created a quite unique stance when it comes to theorizing agency in our discipline.

Using the example of the city as a work domain, Rasmussen, Pejtersen and Goodstein distinguish between functionality and intentionality in a means-ends abstraction hierarchy (1994, p. 43). In their understanding, the functional and material features of the work system dominate the representation at the lower levels, while the intentional features, that is, the objectives that govern the control of the system functions, dominate at the higher levels. They underline the importance of the intentional part as a coordinative force on all levels of the material world: "The intentional component becomes increasingly influential at the higher levels and also more complex as global functions come into play. It can also be seen [in the example of the city] that the intentional part of the domain representation constitutes a hierarchical control function that serves to coordinate the behavior of the material world at all levels (Rasmussen et al., 1994, p. 44)."

Intentions require agency that is exerted by an individual or group. Intentions are based on choice and they can (at least potentially) be communicated. That mostly applies for situations in which a predefined solution, routine or action path is not available. Intentions are the result of an analytic and anticipatory thought process about the control opportunities at hand. According to Rasmussen, Pejtersen and Goodstein, the analysis of the hierarchical control function or intentional part of a work system - becomes more important in systems with sources of regularity other than physical processes (e.g. in manufacturing that involves manual work, companies that provide services, or design) to understand the work domain.

However, when attempting to analyze the cognitive strategies that are involved, the missing conceptual definition of 'intentionality' becomes obvious. Two conceptually interrelated cognitive activities are mentioned by Rasmussen and colleagues, 'functional implications' and 'intentional explications'. Functional implications could be understood as the result of a cognitive process comparing the actual state of affairs with required (or intended) outcomes on different levels of the abstraction hierarchy, providing a link between the 'functional domain' and the 'intentional domain'. Intentional explications on the other side are the result of a cognitive process that dynamically updates intentional states (derived from 'ultimate goals') in order to lead or guide activities on within the 'functional domain' - an therefore requiring explicit communication of corresponding intentions (e.g. through tactical orders, job sequences or schedules).

Intentional explication is shaping individual decision making. But the ontological status of the intentional structure of a work system remains somewhat underspecified: "(...) the actual, individual goals of the smaller units are not found by a decomposition of the overall goal but are developed independently from subjective preferences. Basically, this is a consequence of some of the intentional structure of a work system being embedded in the rules of conduct of the social system and some of it being brought to bear by the individual actors in order to resolve the remaining degrees of freedom (Rasmussen et al., 1994, pp. 45-46)."

To describe a work domain - including its intentional side - it is therefore necessary to chose a set of suitable methods in order to cover both structural aspects, (a) rules of conduct and (b) individual preferences. But then, to operationalize and clarify the theoretical concepts, the terminology is used in a somewhat vague way. What is meant by 'rules of conduct'? And, are the individual preferences really due to personality traits (like risk aversion) or rather an expression of professional knowledge and experience-based interpretations of the state of affairs? - The clarification of these issues is important, since neither role identities nor personality traits are within the scope of a work domain analysis in Rasmussen and colleague's framework. Assuming that both features of the intentional structure can be formulated in a de-personalized way, a cognitive engineering approach to support control decision-making activities must consider the adequate level of representation in both aspects of work domains (cf. Rasmussen et al., 1994, p. 48ff).

My analysis of local rationalities and their modeling using the i* method in case study one can serve as an example of how these intentional structures could be elicited and made explicit, without referring to individual persons. The intentionality of groups or roles in the organization becomes more visible through the description of soft goals and strategic dependencies. Despite these achievements, there remain some open questions. One of them is how intentional structures translate to actual strategic and tactical decisions within day-to-day planning and scheduling activities? Another question targets the complications stemming from historical factors, such as trust and alliances. What can be expected from the other actors at a specific point in time? Reliability or suspicion could be strong constraints for decision-making in an organization.

Control in organizations is distributed among many actors. Some of them do have the powers to change intentional structures – for example by advocating their 'materialization' in defined rules, standards, or in the form of technological artifacts. Case study two has provided some insights into the workings and dynamics of conditioning and inscription as well as interaction and modification of social structures. These structures, material as well as non-material, are forming the overall constraints for agency in organizations. Their heterogeneity complicates things further: Some rules are explicit, others are not. Some persons have access to information and guidance, others have not. And some members of the organization are in positions that support networking, whereas others are isolated.

Rasmussen and colleagues are referring to informal as well as formal rules of conduct. Formal rules could be understood to consist of communicated standards or procedures, and even routines. Informal rules of conduct might be non-documented routines or behavioral repertoires that have emerged within a community of practice over time (Wenger, 1999), or they might be rooted in discursive norm circles (Elder-Vass, 2012a). In general, work systems that are relatively loosely coupled require more active explication, communication and interpretation of intentional information, i.e. more intentional activities on all levels. As Rasmussen and colleagues summarize, "coordination and control of activities [in loosely coupled systems] depend on the communication of company-institutional objectives. The intentionality originating from the interpretation of environmental conditions and constraints by the management (...) propagates dynamically downward and becomes implemented in more detailed policies and practices by members of the staff. Making intentions operational and explicit during this process requires an interpretation considering a multitude of details dictated by the local context (Rasmussen et al., 1994, pp. 51-52)."

As a consequence, some members of an organization are charged with the 'translation' of management intentionality (as 'interpretations of environmental conditions') into policies and practical directions for workers further below in the hierarchy. They need to resolve many degrees of freedom not only for themselves, but for many others as well. This in turn implies that the individual actor faces a work environment in which the regularity to a considerable degree depends on the intentionality brought to bear by colleagues. The question of how others might react adds considerable complexity to the problem solving and decision making tasks of these persons. But they – due to their somewhat privileged position and usually also access to higher levels of the hierarchy – also have various possibilities to actively change the intentional structure throughout the organization.

Agency in organization therefore is unevenly distributed. Some members have more access to information and higher level 'intentionality' such as interpretations of external events, directives or strategies. These persons are directing their control behavior not only at the functional state of the production process, but also on other actors' intentions. Thirdly, they might even consider to modify the intentional structure embedded in the control system of the organization (cf. Rasmussen et al., 1994, p. 52f). In an environment with predominantly direct face-to-face interaction, intentions can be made explicit and aligned through a process of negotiation or mediation as conflicting intentions become evident. Within a distributed work environment involving computers as collaborative tools, these dynamics within the 'intentional domain' are hindered or obscured substantially.

Technology therefore plays an important role both in restricting and enhancing agency. Depending on the details of the implementation, intentions become more or less explicitly 'readable' by the users of the system. Potentially, strategies and tactical decisions might even be made transparent and documented for other actors. Usually this is not the case – but experienced users are able to 'read the game' of their colleagues within the different layers above and below them. Yet, some of the intentional aspects that affect their collaboration might never become evident, and are constraining the overall performance of the system without being addressed in the organization. But how can such constraints be represented? How can we formalize intentions on various levels of organizations? And, who needs to know about them? Accordingly, at the end of their book, Rasmussen and colleagues add an outlook to their conceptual elaborations on intentionality: "We have not been able to find 'ecological' displays for analysis and evaluation of work system states for autonomous system users in a constrained environment (...), that is, for work domains in which the source of regularity is related to intentional structures, such as laws, regulations, and company strategies. The reason for this situation is that symbolic representations of intentional structures have not as yet been developed. Activities, such as production flow control in just-in-time production systems, patient treatment planning in hospitals, or case handling in public service institutions, are handled by autonomous actors within the constraints posed by regulations, by the intended product-case flow, and by the intentions of colleagues handling the state of affairs up- and down-stream of the flow of work items. For design of ecological interfaces, we need to formalize a description of the different forms of intentional structures and to find suitable symbolic representations. This is an area for further development (Rasmussen et al., 1994, p. 330)."

When relating to today's ERP systems, these – more or less autonomous – actors are not necessarily privileged users of an open and friendly system. They are assigned predefined roles and rights, and they get trained as how to use the system in the prescribed way. As a necessary but also restraining condition of their professional existence, is hardly possible for them to use it 'with heart' (Nardi & O'Day, 2000). The ERP environment with its overwhelming dimensions resembles more to what Ellul (1988) has called a 'technique' (a kind of technological network or complex). The 'raison d'être' (right or reason to exist) for this interconnected 'technique' is to formalize and structure the production process according to certain principles, mainly productivity. Or, as Webster formulated: "If technological development has a logic, it is not autonomous logic, but one which arises from attempts by corporate managements to organize production to meet their particular objectives (Webster, 1991a, p. 199)." When analyzing technology use in organizations, we as psychologists and engineers should not forget these fundamental preconditions.

It is an important step for Human Factors to understand and theoretically frame agency in a more comprehensive way. Agency is directed towards possibilities, it thus requires information about the state of affairs as well as degrees of freedom to act upon them. Agency is always already restricted in organizations, it is a property of a position or role. Technology acts as an opportunity or constraint for individuals and groups to enact their agential powers in specific situations. Implementing a technology usually brings new opportunities as well as risks for the intentional structure of an organization. Some agents loose agency, others are gaining in the process. In general, agency is unevenly distributed and agents that are striving to attain their goals are inclined to come up with way to boost their agential powers. That could for example be achieved through informal relationships, professional networks or even 'hacking' of the IT system.

Agency seen from a systems perspective has – to my knowledge of literature – not received sufficient scholarly attention in Human Factors. This is partly due to a lack of ontological clarity as to what exactly agency is and how it is embedded in social structures of organizations. A critical theory/methods package for Human Factors must address these issues in one way or the other.

7.3 Critical realist theory/methods package

The theory/methods package I have developed in this thesis using the work domain of production planning and scheduling as a practical background is built upon a realist social ontology. More specifically, a local ontology for a discipline like Human Factors must account for the particular *types of entities* that constitute the objects of the discipline, the *parts* of each type of entity and the *sets of relations* between them that are required to constitute them into this type of entity, the *emergent properties* or causal powers of each type of entity, the *mechanisms* through which their parts and the characteristic relations between them produce the emergent properties of the wholes, the *morphogenetic causes* that bring each type of entity into existence, the *morphostatic causes* that sustain their existence, and the ways that these sorts of entities, with these properties, interact to cause the *events* we seek to explain (Elder-Vass, 2010, p. 69).

The crucial types of entities for Human Factors are professional *individuals*, working in *organizations* with a certain division of labor, routines, internal systems of regulation and control and a technological *infrastructure* (artifacts).

Organizations are structured social groups with emergent causal powers. They depend on normative mechanisms to produce the *role specialization* upon which they depend, but *role-coordinated interaction* between their members (which may include non-human material things) provides a further class of mechanisms, a class that confers non-normative causal powers on the organizations concerned (Elder-Vass, 2010, p. 167). Authority relations are a variety of role specialization, they confer some part of the power of the organization as a whole on certain persons (as role occupants). The management role includes the development of the role specifications in response to the goals, performance and circumstances of the organization.

Infrastructure consists of technological systems that have properties and powers analog to social structures, for example though the distribution of user roles and access rights to locations, functions, data or content. Technological systems are physical as well as 'institutional' in the sense that they are social structures of a material kind. They embody norms of behavior, degrees of freedom or constraints to agency that have been inscribed to them by their designers, who in turn have been influenced by their own social environment. Their emergent powers stem from a combination or constellation of embedded resources like algorithms that access large databases with layers of data presentation and inscribed access and manipulation potentials.

A regional theory of a technological system must therefore be concerned with identifying the causal mechanisms underlying emergent properties, and with explaining how these interact to produce events of interest (cf. Elder-Vass, 2010). The identification of causal mechanisms (retroduction) requires the use of established methods for empirical research, including both quantitative and qualitative techniques. Therefore I have proposed to use constructivist social theory (in the 'weak' sense of constructivism, cf. Elder-Vass, 2012a) and a set of qualitative methods to approach and elicit social structures like discursive norm circles and other underlying properties of organizational actors.

In realist social ontology, agency and structure are depending on each other, but are *analytically* treated as separate entities or aspects. On a *philosophical* ground, given their co-constitutive nature, the dualism can be contested with good reasons, as do proponents of Gidden's structuration theory or its modern derivatives like the sociomateriality framework (e.g. Caldwell, 2007; Leonardi, 2013; Orlikowski & Scott, 2008). But, on an *analytical* ground, there are strong arguments for such an *analytical dualism* (e.g. Archer, 1995; Mutch, 2013). Mainly for the reason of its potential for critique, I chose the second approach for my proposed theory/methods package. In separating agency and structure, the analysis of work processes becomes not only more intelligible, but also creates more points of leverage for design and resistance.

Further, a dual approach allows for the detailed description of hybrid socio-technological constellations of action and the role of local rationalities/knowledge in the production of actual events. It avoids treating heterogeneous actors and actants as 'equal' participants and it makes a historical or morphogenetic explanation of interactions and subsequent modifications possible through acknowledging the role of temporal order. In doing so, agency embedded in non-human structures and artifacts can be theoretically conceived and become practical relevance in system design (cf. Rammert, 2007).

In taking this approach, we can develop 'mid-range' theories about technology. An example of a mid-range theory hat is possible on such philosophical assumptions is Normalization Process Theory. This theory provides a propositional framework about processes of implementation of 'material practices', and the work that actors do to achieve it (May & Finch, 2009). According to May and Finch, "Normalization Process Theory provides a robust and replicable ecological framework for analyzing the dynamic collective work and relationships involved in the implementation and social shaping of practices. It is a theory for empirical application rather than abstract critique. It focuses attention on organized and organizing agency in the production and reproduction of the implementation, embedding (or not), and continuing integration of material practices (May & Finch, 2009, p. 549)".

It has not been the objective of this thesis to come up with new theories about production planning and scheduling as a practice. My goal was to propose a coherent and useful theory/methods package for a critical approach to Human Factors. The package should be capable of critique in the sense of sensitivity to workings and embedded structures of power as well as self-reflexivity about scientific practice. Ideally, it should allow for 'two-sided' psychological categories, as Holzkamp has claimed for a Critical Psychology: "... if critical psychologists want to avoid becoming hopelessly mired in semantic debate, they will first of all have to have at their disposal an empirical method for the critique and development of categories, since only by such a means can the categories be rescued from the total indeterminancy of semantical arrangements, and, further, they will have to be able to demonstrate with such a method that their 'two-sided' psychological categories have a greater adequacy with regard to the object than the 'one-sided' categories (Holzkamp, 1992, p. 194)."

In order to come up with the final conclusions of this thesis, I will attempt to reflect on the chosen path and material in a self-critical way. Then I will discuss possible remedies for the weaknesses or problem zones that I have been able to come up with. This should allow for concluding remarks about the applicability, validity, usefulness and general scientific value of my contribution.

The first critical self-reflection is about language, since it seems crucial in getting away from essentialist perspectives on technical artifacts and systems. As Grint and Woolgar put it: "As prisoners of the conventions of language and representation, we display, reaffirm and sustain the basic premises of essentialism that entities of all kinds, but most visibly and consequentially technical artifacts and technological systems, possess characteristics and capacities, and are capable of effects. (...) It follows that a radical move away from essentialism (...) requires nothing short of a major reworking of the categories and conventions of language use (1997, p. 114)".

Based on the substantial ground work done by Rasmussen and colleagues (Rasmussen et al., 1994; Rasmussen et al., 1990) as well as Vicente (1999) on a *taxonomy* for cognitive work analysis, I have proposed a realist social ontology to guide further explorations in that direction. This has allowed for some theoretical and methodological extensions – mainly in the analysis of organizational structures – and the integration of the terminology that comes with them. Through the proposed use of socio-cognitive discourse analysis and the morphogenetic approach within the framework of Cognitive Work Analysis, I have introduced categories and language conventions that come from a tradition of scholarly critique and historical-political awareness and sensitivity. In doing so I realized that there is a general compatibility of these concepts with the already existing framework.

While ontology and taxonomy are important if not essential fundaments for research in complex systems, they do not provide answers for all methodological questions. How can we, as psychologists or engineers, address issues of agency, social structure, positioning and discourse to explain success or failure of technical systems? Either we chose not to address these issues at all and restrict ourselves to work domains where agency and social structure are restricted through highly formalized organizational structures (e.g. aviation, specialized control rooms). Or we have to open up for a truly interdisciplinary approach in order to investigate work domains where agency and social structure have a major influence on the overall functioning.

Interdisciplinarity, however, can be understood in at least two ways. The first concept of interdisciplinarity is aiming at a unification of theories and concepts as well as methods. The second concept of interdisciplinarity is aiming at an integration of results without the goal of unification (Danermark, 2002). This approach acknowledges the fact that reality is stratified and that methods have to be chosen according to the strata of interest (e.g. experimental research for biological strata). For the second approach, it is necessary to discuss ontological questions, since it will not suffice to discuss methods alone. All participants in this kind of interdisciplinary research will have to subscribe to the same ontology, otherwise the integration of the results will be impossible. I am suggesting that this is the right way to proceed for developing a critical Human Factors approach that inherently must be interdisciplinary.

According to Danermark, "interdisciplinary research is the study of a common complex phenomenon and how that phenomenon is manifested at different levels of reality. This is done by using specific theories and methods developed for each level. The results are then integrated in an attempt to reach a more holistic perspective on the phenomenon (2002, p. 61)". And further, "interdisciplinary research has to be characterized by methodological pluralism and not by methodological imperialism or methodological relativism. Even so, true interdisciplinary research demands respect for the methodology of the different disciplines (2002, p. 63)."

The theory/methods package that I have proposed above not only allows for an interdisciplinary investigation of complex and distributed sociotechnical work settings but also for a self-reflexive research mode that has the potential to avoid 'single-sided' blind spots and pitfalls. Pollock and Williams have taken a comparable route which they call the Biography of Artefacts Framework: "The project was inspired by our unhappiness with the way that most existing research into packaged software and ERP in particular was framed. As we noted (...) there is a huge literature addressing ERP and the development and adoption of workplace technologies more generally. This research is often weaker in theoretical and conceptual terms than STS (which only constitutes a small share of this literature), particularly in its understanding of innovation, and is less concerned to consider issues of methodology and epistemology. The bulk of these studies are framed, somewhat unreflexively, within particular well-established modes of research, constrained within particular loci, timeframes, disciplinary perspectives and concerns (Pollock & Williams, 2009, p. 81)."

Specifically the combination of a critical realist social ontology with the multimodal framework of Cognitive Work Analysis has shown the potential to lead towards a critical and emancipatory approach to work analysis in Human Factors. Therefore, it is necessary to create an openness to multiple perspectives without losing the common ground of a shared ontology. It is only in this way that we can overcome the weaknesses and the limitation of methodologically rigid but scientifically limited modes of research. The opening that is achieved like this makes a broader perspective possible, and Pollock and Williams argue accordingly: "By adopting multiple locales and differing scales of analysis, we aim to address both the fine structures and the broader structures at play in the emergence and evolution of technological fields, combining the respective advantages of local and larger-scale analysis, and in so doing overcoming local-global and action-structure dichotomies (2009, p. 278)."

The general Human Factors problems and issues I have described in chapter 3.3 for the field of production planning and scheduling seem to be addressable with such an approach. Table 21 is providing a schematic comparison of the two approaches discussed here, the conventional Human Factors approaches and the proposed approach that could be roughly described as critical-emancipatory.

Conventional Human Factors approaches to highly intentional work domains remain insufficient due to a subjectivity-free, positivist ontology and methodology that is ignoring power structures and ideologies which are shaping social discourses, individual cognitions and artifacts (Dery et al., 2006; Light & Wagner, 2006; Shepherd et al., 2009). They tend to be blind to the implementation of political agendas through the use of technology (Córdoba, 2007; Koch & Buhl, 2001). Besides conceptualizing and modeling human agents as subjects that are subjected to multi-faceted social environments, a critical-emancipatory approach also addresses the specific managerial practice that is inscribed within the ERP system. In such an approach, the ERP system's ability to shape user's activities is theorized as a kind of agency. Whereas human agency is based on

Overall discussion and conclusions 203

intentionality, such non-human agency is based on *affordance* (Ignatiadis & Nandhakumar, 2006).

Table 20: Characteristics of conventional Human Factors approaches and the proposed critical theory/methods package

	Conventional approaches	**Critical theory/methods package**
Ontology	Positivist / empiricist	Critical realist / constructivist
Preferred methods	Task analysis, controlled experiments, modeling	Work analysis, discourse analysis, morphogenetic analysis, action research
Domain descriptions	Procedural	Declarative
Subjectivity / intentionality	Rational operator isolated from social context	Self-reflexive actors embedded in layered social reality
Agency	Cognitivist perspective (decision models, heuristics)	Emergentist perspective (human and non-human agents)
Social structure	Mostly outside research focus (some exceptions, e.g. organizational culture)	Highly influential on behavior through norms, discourses and local practices (practice-positions)

The central contribution of this thesis, the proposition and discussion of a realist-constructivist theory/methods package for Human Factors as an interdisciplinary discipline, leads to new questions. For instance, should we invest more effort into a formalized set of methods or framework for our research? Could Cognitive Work Analysis serve as a vehicle for this, thus becoming more than a collection of complementary methods to analyze work in complex sociotechnical systems, i.e. a pragmatic combination of task and system based approaches (Ernst, Jamieson & Mylopoulos, 2006; Jamieson, 2003) for cognitive engineering and software design? Could this new, comprehensive package be formulated as a truly interdisciplinary framework that has the potential to address future research and development problems in Human Factors domains? Does this approach eventually contribute to a stronger influence of Human Factors in societal change in general (cf. Vicente, 2008)?

Although I have attempted to put the proposed theory/methods package to work in two case studies from industry, its usefulness for research and development still needs to be demonstrated in future contributions. I will dedicate the last chapter of my thesis to discuss some practical implications of my preliminary and in many aspects unfinished work. In doing so, I am particularly interested in questions revolving around *ethical issues* related to Human Factors, around *democratization* of technology governance, around *participative or action research* as a mode of engagement with those real world problems that do not award us to take a neutral stance as investigators.

8 Implications and outlook

> Real change will come not when we turn away from technology toward meaning, but when we recognize the nature of our subordinate position in the technical systems that enroll us, and begin to intervene in the design process in the defense of the conditions of a meaningful life and a livable environment (Feenberg, 1999, p. xiv).

The contribution of this thesis has implications on at least three aspects of Human Factors research and development efforts related to large-scale enterprise IT systems. Firstly, on the theoretical level, it proposes an explicit philosophical orientation based on an critical realist social ontology. Secondly, on the level of socio-technological development projects, it is showing ways to improve stakeholder involvement at early stages of the process. Thirdly, it provides a framework for teaching and training by setting the stage for case studies or other forms of interdisciplinary learning.

In the final chapter I will try to provide an brief evaluation of these implications as well as a set of possible future directions for research on the 'politics of the artifact' (Winner, 1980), general principles of sociotechnical design and the application of the above propositions to the field of ERP and possibly other, comparable fields.

8.1 Theoretical implications

One of the ethical issues linked to my reflections in this thesis is: In whose interest are we - as Human Factors engineers and psychologists - doing our work? Are we committed to contributing towards more humane, decent workplaces (as we prefer to see ourselves)? Or are we trouble shooters working in the interest of those in power to fix whatever the flaws are within their control structure that has been developed to serve there (exploitative) interests, regardless the well-being of the workforce they employ? Are we conscious of our societal role in general? How do we account for ethical and moral principles in our work as public-funded scientists and scholars (Vicente, 2008)?

Scott (1987) is referring to the concept of *dominant coalitions* to answer the question about *who* is setting organizational goals. To me, the notion of coalitions within organizations, with more or less power, makes perfect sense. It is compatible with the critical realist understanding of the causal powers of social structure, e.g. through discursive norm circles, professional guilds or other forms of social structures. Through the recognition of dominant coalitions, it becomes clear that although single individuals and groups may impose goals in organizations, but they hardly ever are able to do so on their own. Organizational goals are therefore always something else, distinct from individual and group goals. There are differences in interests among participants that may or may not be resolved through negotiation. There may be conflicting goals – leading to dynamic shifts in the size and composition of the dominant coalition (cf. Scott, 1987, p. 271). Dominant coalitions could be understood as situationally aligned position-practices that generate power through their mutual commitment to work towards a common goal. These position-practices may under certain circumstances also be those of managers and researchers.

Thus, when we as researchers 'enter the stage' in an organization, regardless of the level within the hierarchy, we become part of the coalition struggle around organizational powers. We do so by defining, interpreting, formalizing and categorizing aspects of organizational reality. As Law puts it: "There is no innocence. But to the extent social science conceals its performativity from itself it is pretending to an innocence that it cannot have. And to the extent that it enacts methods that look for or assume certain structural stabilities, it enacts those stabilities while interfering with other realities (...). We have suggested that the issue is one of 'ontological politics'. If methods are not innocent then they are also political. They help to make realities. But the question is: which realities? Which do we want to help to make more real, and which less real? How do we want to interfere (because interfere we will, one way or another) (Law & Urry, 2004, p. 405)?"

As a consequence, we have to ask ourselves, which are the realities we help to create? According to Winter (2009) the ultimate objective of research should be the empowerment of not only the researcher but also the participants within the research process. The process should lead to a democratization of power and knowledge. In

Feenberg's terms, this is only possible through a critical perspective refraining from technological determinism (Feenberg, 1999). The process of technological development is essentially a social process that – in principle – has the potential to be conducted or mediated in a democratic way. Feenberg describes this mode of development as "democratic rationalization (1999, p. 74ff)". In an analogue vein, Flores and colleagues argue that such processes can be conducted with awareness of the implications that are connected with technological change: "(...) all innovative technology leads to new practices, which cause social and organizational changes whether anticipated or not. Some of these will be effective and others may be counterproductive. Our firm belief is that this process can be done with awareness. Although we can never fully anticipate the changes a technology will trigger, we can make conscious choices in the directions of change we facilitate (Flores, Graves, Hartfield & Winograd, 1988, p. 169)." However, this rather optimist perspective of the creative powers of designers has also been criticized, for example by Suchman, for not accounting for resistance and subversion as legitimate modes of action for certain participants in organizations or societies (cf. Suchman, 1994).

One way of theoretically framing the processes of power and knowledge in technological design is to postulate the communicative nature of any kind of artifact. Technology, especially today's networked ICT infrastructures can be understood as communicative ensembles, involving specific codes, forms, discourses and cultures. They are creating rooms or spaces of shared intentionality through communicative acts, and thus communication power (Castells, 2009; Knoblauch, 2013). Coming from this theoretical perspective, any design process could be navigated into two directions. Either, it could be directed towards greater transparency and translucence to enhance shared intentionality, or, it could be directed towards more obscure and mysterious mechanisms hidden in some kind of 'black box'. Most likely, the decision which direction is most appropriate is taken by management. Depending on the goals and purposes of an organization, the decision makers may chose one way or the other, with more or less awareness of the (social) consequences of their decision.

A hypothetical case could be that an organization, for example a university or a staff-owned not-for-profit service provider, is intending to develop a technological design that leads to the highest possible transparency when it comes to intentionality. One way to achieve this could be through the application of methods of participatory (action) research (cf. Bergold & Thomas, 2012). During the field work conducted to collect the data for our case studies described further above, I have experienced several occasions where we were involved in a process that could be described as a mutual learning experience. At least once we were surprised about the consequences of our own questioning. In a truly participatory research approach, not only the goals of the research but also the methods and samples are defined by both the researchers as well as the participants. As a result, not only the researcher learns about what kind of questions could be asked, but also the participants are introduced into the use of previously unknown methods and instruments.

8.2 Practical implications for research and development

A future elaboration of the ideas and concepts that I developed in my thesis might thus be the participatory reformulation or extension of this approach to complex sociotechnical systems (cf. Bergold & Thomas, 2012). This might be especially useful for example in academia, public administration, non-profit sectors or community-owned Internet platforms. When it comes to ERP implementations, many of these organizations or institutions are struggling with the standardized offers that come with substantial tailoring efforts. These efforts are costly and – because of most IT partners being trained and accustomed to industrial business process modeling and the like – are often not leading to the desired outcomes.

But even when an organization or its management is willing to do its utmost to create an open, democratic process that involves participatory design methods, there still might be resistance, or even sabotage. The health and prosperity, or even the existence of the organization is at stake. So for us as Human Factors experts, the question is: What are our concerns and strategies when designing systems that are used in an organizational environment including users that follow different, and possibly opposed objectives or policies?

Some critical psychologists would make these voices or needs the starting point of their work. While this certainly makes sense from an ethical and social perspective, it might also make sense for an

Implications and outlook

organization in economical dimensions, especially on the long run. One example are training efforts. It makes a big difference if training participants think or believe the technology is degrading or harmful, or if they believe it will improve their quality of work-life. As Martín-Baró argues, "the political impact that can result from the psychologist's work in the company, whether direct or indirect, deserves special emphasis. To take one example, the psychologist cannot define jobs according to the internal logic of the institution and the particular characteristics of individuals, while ignoring contributing criteria in a country's social, political, and cultural spheres; for if these are neglected, what functions as adaptation to the business risks becoming a variety of political submission, and what appears on the surface as nothing more than technical training bears within it the seed of political and social alienation. Thus, when labor psychology fails to make empowerment one of its goals, or the active and organized participation of workers one of its principles, it runs the danger of falling into psychologizing and promoting alienation even in the act of disguising it (Martín-Baró, 1994, p. 98)."

In following a more pragmatic approach, but one compatible with the considerations laid out above, Marmaras and Nathanael (2005) are proposing to systematize the reality of cognitive engineering practice and put this framework to use in concrete design processes. In their view, it is especially important to (1) explicitly consider the demand for intervention and the analysts preconceptions during the framing of the world-to-study, to (2) adopt multiple views for a sufficient understanding of the world-of-practice, to (3) continuously reframe the world-to-study (i.e., topological boundaries as well as timeframes) as the understanding of reality unfolds, and to (4) accept and exploit the dialectic process between analysis/understanding and design/prediction (Marmaras & Nathanael, 2005, p. 125).

But doing so leads us away from our conventional design strategies: "Admitting that we can only hint at how the future work ecology (artifacts included) will behave, brings us both closer to reality and to a highly disadvantageous position. The cognitive engineer must accept this reality and alter his design strategy accordingly: rather than fully optimising for a partially predicted world, he should consciously expect the turbulence of reality. Therefore, he/she should concentrate not on accuracy or pure optimisation, but on ensuring against undesirable outcomes and specify features of the artefact in a way that it will be better suited for adaptation (Marmaras & Nathanael, 2005, p. 124)."

In following this path, system design becomes more critical towards promises of perfect optimization, more oriented towards adaptability and resilience, to shifts and transformations in intentionalities and organizational politics. It becomes of crucial importance, for example, *who* defines the process by which requirements are established (cf. Ernst et al., 2006). Only through a careful consideration of all aspects of the world-of-practice, we can come up with a more complete analysis as a base for successful design. This requires, among other things, that we refrain from conceptualizing an organization as a system in a narrow sense. Feenberg therefore proposes to use the term *network* instead of system. As he puts it, "the intentions of managers are no more fundamental than the vagaries of people (and things) enrolled unintentionally in the network of which the 'system' is a subset. A network theory of the technical politics in which these unofficial actors engage needs new categories that do not depend on the self-understanding of managers (Feenberg, 1999, p. 119)."

In following Feenberg, the design of artifacts turns into an aspect of technical politics within the organizational network. That system design is involving political processes is nothing new, but is not widely discussed in the design nor in the Human Factors community. In his comprehensive paper on sociotechnical principles for system design, Clegg (2000) theorizes the design process not only as an extended social process, but also as inherently political. According to Clegg, "there need to be mechanisms in place to handle these political debates and discussions in ways acceptable to those affected by the designs (Clegg, 2000, p. 474)". Clegg furthermore stresses the importance of evaluation, hereby including social, technical, operational and financial aspects. Accordingly, it would make sense to promote project structures involving managers, users, technology providers as well as independent evaluators. The evaluation process itself should be designed by all relevant stakeholders involved.

Political negotiations and formative as well as summative evaluations within sociotechnical design processes could be understood as a form of organizational learning. Learning only becomes possible through established feedback mechanisms. Besides the evaluation of technological change and its organizational impacts in specific projects, there is a need for the creation of more general learning spaces that are more independent of producers and their associated vendors and consultancies. To establish such spaces, the support of research agencies might be necessary: "Governmental technology policy could (...) promote a more experimental learning oriented approach by funding a more systematic collection of experiences and in this way support the development of understandings and concepts, facilitating the exchange of experiences and the social shaping of socio-technical ensembles of CAPM and ERP (cf. Clausen & Koch, 1999, p. 481)".

8.3 Implications for teaching and training

Within Organization Sciences, the calls for a more interdisciplinary approach on a comprehensive theoretical base are getting stronger as well. Volkoff and colleagues are proposing a theory of technology-mediated organizational change because understanding technology-mediated organizational change is becoming a required skill for successful managers. They suggest that managers need to consider more than just the planned changes to data and functionality. Through a systemic understanding of large-scale enterprise IT systems organizational decision makers are getting more competent in guiding the change processes involved. Arguing for their own theoretical approach, Volkoff and colleagues state: "Such systems are likely to affect a broader range of organizational elements, such as roles, relations between data and routines or routines and roles, forms of control and mindset. In addition, our theory helps managers understand how organizational elements differ in their changeability and their time cycle for changing (Volkoff et al., 2007, p. 846)."

This view is in correspondence with what Ciborra and Hanseth have observed and described some years earlier. They wrote: "Technology becomes hard to change as successful changes need to be compatible with the installed base. As the number of users grows, reaching agreement about new features as well as coordinating transitions becomes increasingly difficult. Vendors develop products implementing a standard, new technologies are built on top of it. As the installed base grows, institutions like standardization bodies are established, and the interests vested in the technology expand. It follows that designing and governing an infrastructure differs from designing an MIS, due to the far-reaching influence of the installed base and the self-reinforcing mechanisms pointed out by the economists. The very scope of the management agenda should change. Infrastructure is not just a complex, shared tool that management are free to align according to their strategy. The economic perspective highlights a much more limited and opportunistic agenda involving trade-offs and dilemmas, and a number of tactics (Ciborra & Hanseth, 1998, p. 311)."

According to Dahlbom and Mathiassen (1997), one can distinguish three different types of information professionals, depending on their understanding of their role, focus and approach when working on IT projects (engineer/programmer, facilitator/supporter, emancipator/consultant). In their view, it is essential for the future of the profession that these three perspectives become more interlinked, and that information professionals and systems developers are educated with a more coordinated approach to different types of knowledge, including the philosophical dimension of technology and information.

Wenger, White and Smith (2009) have proposed to use the term technology stewardship to designate a novel kind of professional role related to (information) technologies. According to them, technology stewardship is both a perspective and a practice that can be considered as a collection of activities carried out by individuals within a community. They write: "Technology stewards are people with enough experience of the workings of a community to understand its technology needs, and enough experience with or interest in technology to take leadership in addressing those needs. Stewarding typically includes selecting and configuring technology, as well as supporting its use in the practice of a community (Wenger et al., 2009, p. 25)."

Discussing theoretical and practical aspects of agency and social structures in relation to large-scale information infrastructures might help to educate and coach managers, IT professionals as well as others who are concerned with the implementation and use of such systems in a dynamic and networked environment.

8.4 Possible future directions

Where could we go from here? Since Human Factors is about the improvement of human-technology relations, the most obvious path would be to continue working on better analytic methods and design concepts. This could for example be achieved by extending and reformulating parts of the Cognitive Work Analysis framework. While this certainly provides an interesting road which is worthwhile exploring, I will focus my discussion on future directions on less obvious pathways to travel.

In an edited book about virtual social interaction and navigation, Chalmers is linking the design of 'information spaces' to architecture theory. This allows for interesting parallels, not only theoretical but also practical in nature. While Chalmer's focus is on semiotics and the related discussion of theories of information and architecture, his discussion could well be taken as a starting point of a more dialectical approach to information systems design. For instance, the notion of city and its continuous appropriation by the citizens who live and work in this architectural space can be used as a kind of metaphor for large-scale networked information spaces such as collaborative software or enterprise systems. Users are creating shortcuts where designers did not think of providing direct access. Others are meeting in places that were not designed for social happenings. Just as public space becomes more and more contested between private (corporate) interests and the inhabitants and visitors of a city, information spaces might also be understood as 'contested' areas or spheres, where users are voicing their concerns and needs to shape these places in favor of their interests (Chalmers, 2003).

An interesting but mostly unknown Human Factors design approach is Community Ergonomics (an overview can be found in: Smith et al., 2002). While this approach initially was developed for distressed community settings characterized by poverty, social isolation, dependency, and low levels of self-regulation, it could possibly be translated and adapted to virtual communities using large-scale information systems. Interestingly, transparency about other user's actions is an important topic in Community Ergonomics. A very similar demand was formulated elsewhere, calling for 'social translucence' of social media and collaboration platforms (Höök, Benyon & Munro, 2003). Participative design, control and self-regulation are important aspects identified by Community

Ergonomics. Consequently, the principles that drive Community Ergonomics are (1) an action-oriented approach, (2) participation by everyone, (3) diversity and conflict management, (4) encouraging learning, (5) building self-regulation, (6) feedback triad, and (7) continuous improvement and innovation.

Seen through the lens of the community, technological change is a cultural process that can be potentially be influenced and controlled by the members of the community. A technology needs to be cultivated in a sense, put into a context and embedded into a practice that is always pre-dating the technology. Vice versa, the technology is sending out its own cultural influence and is therefore acting as a catalyzer for change within the community. Both aspects need to be considered and have not yet been thoroughly discussed within Human Factors. One of the problems of 'traditional' Socio-Technical Design and Macroergonomics is their inherent top-down approach (cf. Hendrick & Kleiner, 2002). Members of an organization, as a community, do not just have to be organized in a specific way to fit to the (technological) environment to solve problems related to safety and reliability. They also need to be taken seriously as a cultivated, civilized collective of groups and individuals. Much like becoming a member of an organization is a process of enculturation, embedding a technology into a community is a process of enculturation and assimilation.

Some very important and not well researched aspects of human behavior in complex systems are mindfulness, organizational citizenship and democratization of information technology. The question of mindfulness is especially relevant in large-scale technical systems or high reliability organizations: "Simplifications produce blind spots (Weick & Sutcliffe, 2001, p. 62)". To achieve high levels of mindfulness, it is necessary that people know and respect each other. Trustful relationships facilitate not only open discussions of known issues and potential problems, they also enhance what has been called organizational citizenship behavior (Boiral, 2008; Lievens & Anseel, 2004). Valorinta (2009) analyses how information technology impacts mindfulness in organizations. He finds that IT-intensive businesses share many features of High Reliability Organizations like nuclear power plants or air traffic control systems. Failure is never an option, and even small errors can lead to catastrophic consequences. Therefore mindfulness, i.e. being constantly attentive and open to

Implications and outlook

unfamiliar interpretations of the environment and responding quickly to early signs of trouble, becomes an important capacity of such organizations. However, the behaviors and processes described by Valorinta are focussing on an organizational or project management level. Apart from the 'unifying language' of ERP systems that facilitates cross-departmental interactions, the users of the system are – so it seems – not 'in the loop'. Valorinta's study concentrates mostly on IT professionals and management and their respective contributions to mindfulness.

Therefore, an interesting applied research question could be if it is possible to design, tailor and implement ERP systems in a way to support and enhance mindfulness in a specific organization *on all levels*. As Pinch puts it, "to understand how technologies enable and constrain social interaction, it is important not to take either their constraining or enabling features for granted and study both how technologies could be different and how social interaction built around technologies could be different (Pinch, 2008, p. 469)". To achieve that, the research design would need to work in parallel on the social structures that 'govern' the technology as well as on features of the technology to allow for the desired mindful practices or work cultures. In this regard, Feenberg's ideas about democratization of technology might provide innovative guidelines to come up with new design and development methods (Feenberg, 1999).

To intensify work in this area it is necessary to apply a methodology backed by an ontology linking individuals with social structures, including technologies, and building on an epistemological framework that serves as a point of departure for empirical studies. The realist-constructivist theory/methods package sketched out in my thesis could possibly serve as an entry or passageway to more theoretical and practical work related to processes of culture and technology in organizations.

One technology that certainly deserves more attention is ERP systems. As Koch (2005) has proposed, research on ERP systems hereby needs to broaden its horizon. It should not only be focused on the implementation or redesign phase, and not only on the micro-level of the IT department struggling with the project. Table 22 is showing the different levels and temporal horizons such research endeavors may cover in the future. Following the above discussion on mindfulness, I have added one more layer to Koch's proposed three layers. The nano layer is targeting changes in the workplace, the tasks and the interactions of the users of the system. On the long term, the changes on this level might lead to a different work culture or changes in local rationalities. One of these changes might be enhanced or impaired mindfulness.

Other large-scale, distributed information systems that are of almost existential importance to organizations can be found in banks and finance organizations, insurance companies, retail industry, transport and telecommunication operators (cf. Valorinta, 2009). However, doing comprehensive research in these contexts probably involves the general rethinking the role of social theory in socio-technical analysis. Pettersen, McDonald and Engen argue that "today social science tends only to play a role in evaluation, training, investigation and is not central to design and development new technologies or new operational concepts. (...) this is caused by a fundamental lack of theoretical basis in current approaches for both developing new and intervening in current socio-technical systems in transport and other industries (Pettersen et al., 2010, p. 190)". The theory/methods package proposed in this thesis might help to open up analytical paths and critical inquiries about the consequences of such technologies as well as ways to improve their design, implementation and governance.

Table 21: Spatiality and temporal spanning of ERP studies (adapted from Koch, 2005, p. 53)

	Short term	**Long term**
Nano	Workplace, task, interaction	Work culture, mindfulness, local rationality
Micro	Purchasing, implementation	Life cycle 'after going live', modifications
Meso	Professional associations, network constellations	Institutions of technology, biography of system elements
Macro	Technology policy, promotion	Global technological and company change, communities

Implications and outlook

A critical Human Factors perspective might not only help to solve issues in organizations, it also has the potential to tackle technology-related problems on a more global scale. There are many of these problems still waiting for interdisciplinary solutions, mainly in the area of sustainability and the protection of the environment (cf. Flemming, Hilliard & Jamieson, 2008). Social structures have a strong influence on policies and societal change or resistance to it. To understand technology in relation to human agency and social structures will help to find ways to better deal with the powerful tools and systems that are available today. From a critical realist perspective, "Human Factors researchers and practitioners will be better able to see how societal problems can rarely be solved by purely technical solutions alone; knowledge of organizational, social, and political forces is essential to understanding and fostering change (Vicente, 2008, p. 22)."

Human Factors needs to develop that kind of knowledge, especially for today's complex organizations which are employing large-scale information technology systems that are increasingly distributed, dynamic and adaptive.

References

Abrahamson, E. & Baumard, P. (2008). What lies behind organizational façades and how organizational façades lie: An untold story of organizational decision making. In G. P. Hodgkinson & W. H. Starbuck (Eds.), *The Oxford Handbook of Organizational Decision Making*. Oxford: Oxford University Press (pp. 437-52).

Ahrens, V. (1998). System Models and Concepts for Distributed Production Planning and Control. In E. Scherer (Ed.), Shop Floor Control: A Systems Perspective. Springer Verlag (pp. 173-97).

Akkerman, R. & van Donk, D. P. (2009). Analyzing scheduling in the food-processing industry: Structure and tasks. Cognition, Technology & Work, 11(3), 215-226.

Akkermans, H. & van Helden, K. (2002). Vicious and virtuous cycles in ERP implementation: a case study of interrelations between critical success factors. European Journal of Information Systems, 11, 35-46.

Alterman, R. (1988). Adaptive planning. Cognitive Science, 12(3), 393-421.

Amoako-Gyampah, K. & Salam, A. F. (2004). An extension of the technology acceptance model in an ERP implementation environment. Information & Management, 41(6), 731-745.

Archer, M. (1998). Introduction: Realism in the social sciences. In M. Archer, R. Bhaskar, A. Collier, T. Lewson, & A. Norrie (Eds.), Critical Realism: Essential Readings. London: Routledge (pp. 189-205).

Archer, M. S. (1995). Realist social theory: The morphogenetic approach. Cambridge: Cambridge University Press.

Archer, M. S. (2002). Realism and the Problem of Agency. Journal of Critical Realism, 5(1), 11-20.

Argyris, C. (1977). Organizational learning and management information systems. Accounting, Organizations and Society, 2(2), 113-123.

Ashby, W. R. (1957). An introduction to cybernetics. London: Chapman & Hall.

Avis, J. (2007). Engeström's version of activity theory: a conservative praxis? Journal of Education and Work, 20(3), 161-177.

Axel, E. (2003). Theoretical Deliberations on "Regulation as Productive Tool Use". Outlines, 5(1), 31-46.

References

Badham, R. & Schallock, B. (1991). Human factors in CIM: A human-centered perspective from Europe. Human Factors and Ergonomics in Manufacturing, 1(2), 121-141.

Badke-Schaub, P., Hofinger, G., & Lauche, K. (2008). Human Factors: Psychologie sicheren Handelns in Risikobranchen. Heidelberg: Springer Verlag.

Barker, R. G. (1968). Ecological Psychology. Stanford CA: Stanford University Press.

Bateson, G. (1972). Steps to an ecology of mind: A revolutionary approach to man's understanding of himself. New York: Ballantine.

Baumann, Z. (1992). Intimations of Postmodernity. London: Routledge.

Bechky, B. A. (2003). Sharing meaning across occupational communities: The transformation of understanding on a production floor. Organization Science, 14(3), 312-330.

Beck, U. (1992). Risk Society: Towards a New Modernity. London: Sage Publications.

Becker, M. C. (2004). Organizational routines: a review of the literature. Industrial and Corporate Change, 13(4), 643-678.

Benders, J., Hoeken, P., Batenburg, R., & Schouteten, R. (2006). First organise, then automate: a modern socio-technical view on ERP systems and teamworking. New Technology, Work and Employment, 21(3), 242-251.

Benz, M. (2007). The relevance of procedural utility for economics. In B. S. Frey & A. Stutzer (Eds.), Economics and Psychology. Cambridge, MA: MIT Press.

Berger, P. L. & Luckmann, T. (1967). The Social Construction of Reality. New York: Anchor Books.

Berglund, M., Guinery, J., & Karltun, J. (2011). The Unsung Contributions of Planners and Schedulers at Production and Sales Interfaces. In J. Fransoo, T. Wäfler, & J. Wilson (Eds.), Behavioral Operations in Planning and Scheduling. Berlin: Springer (pp. 47-82).

Bergold, J. & Thomas, S. (2012). Partizipative Forschungsmethoden: Ein methodischer Ansatz in Bewegung. Forum Qualitative Sozialforschung/Forum: Qualitative Social Research, 13(1), Art. 30.

Betsch, T. (2005). Preference Theory. In T. Betsch & S. Haberstroh (Eds.), The Routines of Decision Making. Mahwah: Lawrence Erlbaum (pp. 39-66).

Betsch, T. & Haberstroh, S. (2005). Current Research in Routine Decision Making: Advances and Prospects. In T. Betsch & S.

Haberstroh (Eds.), The Routines of Decision Making. Mahwah: Lawrence Erlbaum (pp. 359-76).

Betsch, T., Haberstroh, S., & Höhle, C. (2002). Explaining Routinized Decision Making. Theory & Psychology, 12(4), 453-488.

Bhaskar, R. (1975). A realist theory of science. London: Verso.

Bhaskar, R. (1979). The Possibility of Naturalism. London: Routledge.

Bhaskar, R. (1998). Philosophy and scientific realism. In M. Archer, R. Bhaskar, A. Collier, T. Lawson, & A. Norrie (Eds.), Critical Realism: Essential Readings. London: Routledge (pp. 16-47).

Bijker, W. E., Hughes, T. P., & Pinch, T. (1987). The social construction of technological systems. Cambridge MA: MIT Press.

Bina, B., Chen, H. W., & Milgram, P. (2008). A Model of Expert Decision Making in Post-Flop Betting in Poker. In Human Factors and Ergonomics Society 52nd Annual Meeting. New York.

Bisantz, A. M. & Ockerman, J. J. (2002). Informing the evaluation and design of technology in intentional work environments through a focus on artefacts and implicit theories. International Journal of Human-Computer Studies, 56(2), 247-265.

Bloomfield, B. P., Latham, Y., & Vurdubakis, T. (2010). Bodies, Technologies and Action Possibilities. Sociology, 44(3), 415-433.

Blumer, H. (1990). Industrialization as an agent of social change: A critical analysis. New York: De Gruyter.

Boesch, E. E. (1991). Symbolic action theory and cultural psychology. Berlin ; New York: Springer-Verlag.

Boiral, O. (2008). Greening the Corporation Through Organizational Citizenship Behaviors. Journal of Business Ethics, 1-16.

Boudreau, M. C. & Robey, D. (2005). Enacting integrated information technology: A human agency perspective. Organization Science, 16(1), 3-18.

Bourdieu, P. (1998). Practical reason: On the theory of action. Stanford: Stanford University Press.

Böhle, F., Bolte, A., Pfeiffer, S., & Porschen, S. (2008). Kooperation und Kommunikation in dezentralen Organisationen - Wandel von formalem und informellem Handeln. In C. Funken & I. Schulz-Schaeffer (Eds.), Digitalisierung der Arbeitswelt. Wiesbaden: Verlag für Sozialwissenschaften (pp. 93-115).

Burns, C. M. & Hajdukiewicz, J. R. (2004). Ecological Interface Design. Boca Raton: CRC Press.

Button, G. & Sharrock, W. (2009). Studies of Work and the Workplace in HCI: Concepts and Techniques. Synthesis Lectures on Human-Centered Informatics, 2(1), 1-96.

Cadili, S. & Whitley, E. A. (2005). On the Interpretative Flexibility of Hosted ERP Systems. London: London School of Economics. Departement of Information Systems Working Paper Series.

Caldwell, R. (2007). Agency and Change: Re-evaluating Foucault's Legacy. Organization, 14(6), 769-791.

Calisir, F. (2004). The relation of interface usability characteristics, perceived usefulness, and perceived ease of use to end-user satisfaction with enterprise resource planning (ERP) systems. Computers in Human Behavior, 20(4), 505-515.

Castells, M. (2009). Communication Power. Oxford: Oxford University Press.

Cegarra, J. (2004). La gestion de la complexité dans la planification: le cas de l'ordonnancement. Thesis, Paris: Université Paris 8.

Cegarra, J. (2008). A cognitive typology of scheduling situations: A contribution to laboratory and field studies. Theoretical Issues in Ergonomics Science, 9(3), 201-222.

Cegarra, J. & van Wezel, W. (2010a). A Comparison of Task Analysis for Planning and Scheduling . In J. Fransoo, T. Wäfler, & J. Wilson (Eds.), Behavioral Operations in Planning and Scheduling. Berlin: Springer (pp. 323-38).

Cegarra, J. & van Wezel, W. (2010b). A Comparison of Task Analysis for Planning and Scheduling . In J. Fransoo, T. Wäfler, & J. Wilson (Eds.), Behavioral Operations in Planning and Scheduling. Berlin: Springer (pp. 323-38).

Chalmers, M. (2003). Informatics, Architecture and Language. In K. Höök, D. Benyon, & A. J. Munro (Eds.), Designing information spaces : the social navigation approach. London ; New York: Springer (pp. 315-42).

Christoffersen, K. & Woods, D. D. (2002). How to make automated systems team players. Advances in Human Performance and Cognitive Engineering Research, 2, 1-12.

Ciborra, C. U. & Hanseth, O. (1998). From tool to Gestell: Agendas for managinig the information infrastructure. Information Technology & People, 11(4), 305-327.

Ciborra, C. U. & Lanzara, G. F. (1994). Formative Contexts and Information Technology: Understanding the Dynamics of Innovation in Organizations. Accounting, Management and Information Technology, 4(2), 61-86.

Clarke, A. (2005). Situational analysis : grounded theory after the postmodern turn. Thousand Oaks: Sage.

Clases, C. & Wehner, T. (2002). Steps across the border - Cooperation, knowledge production and systems design. Computer Supported Cooperative Work (CSCW), 11(1), 39-54.

Clausen, C. & Koch, C. (1999). The role of spaces and occasions in the transformation of information technologies-Lessons from the social shaping of IT systems for manufacturing in a Danish context. Technology Analysis & Strategic Management, 11(3), 463-482.

Clegg, C. W. (2000). Sociotechnical principles for system design. Applied Ergonomics, 31(5), 463-477.

Clegg, S. & Wilson, F. (1991). Power, technology and flexibility in organizations. In J. Law (Ed.), A Sociology of Monsters - Essays on Power, Technology and Domination. London: Routledge.

Collier, A. (1994). Critical Realism: An Introduction to Roy Bhaskar's Philosophy. London: Verso.

Córdoba, J. R. (2007). Developing inclusion and critical reflection in information systems planning. Organization, 14(6), 909-927.

Crandall, B., Klein, G. A., & Hoffman, R. R. (2006). Working minds : a practitioner's guide to cognitive task analysis . Cambridge: MIT Press.

Crawford, S. (2000). A Field Study of Schedulers in Industry: Understanding their Work, Practices and Performance. Thesis, Nottingham: University of Nottingham.

Crawford, S. (2001). Making sense of scheduling: the realities of scheduling practice in an engineering firm. In B. McCarthy & J. Wilson (Eds.), Human performance in planning and scheduling. London: Taylor & Francis. (pp. 83-104).

Crawford, S., MacCarthy, B., Wilson, J. R., & Vernon, C. (1999). Investigating the work of industrial schedulers through field study. Cognition, Technology & Work, (1), 63-77.

Crowston, K. G. (1991). Towards a coordination cookbook: Recipes for multi-agent action. Thesis, Cambridge, MA: MIT.

Cumbie, B., Jourdan, Z., Peachey, T., Dugo, T., & Craighead, C. (2005). Enterprise resource planning research: where are we now and where should we go from here? Journal of Information Technology Theory and Application, 7(2), 21-37.

Currie, W. (2009). Contextualising the IT artefact: towards a wider research agenda for IS using institutional theory. Information Technology & People, 22(1), 63-77.

Dahlbom, B. & Mathiassen, L. (1993). Computers in context : the philosophy and practice of systems design. Cambridge, MA: NCC Blackwell.

Dahlbom, B. & Mathiassen, L. (1997). The future of our profession. Communications of the ACM, 40(6), 80-89.

Danermark, B. (2002). Interdisciplinary Research and Critical Realism: The Example of Disability Research. Journal of Critical Realism, 5(1), 56-64.

Davenport, T. H. (1998). Putting the enterprise into the enterprise system. Harvard Business Review, 76(4), 121-131.

Debitz, U. (2005). Die Gestaltung von Merkmalen des Arbeitssystems und ihre Auswirkungen auf Beanspruchungsprozesse. Hamburg: Kovac.

Dery, K., Grant, D., Harley, B., & Wright, C. (2006). Work, organisation and Enterprise Resource Planning systems: an alternative research agenda. New Technology, Work and Employment, 21(3), 199-214.

Dessouky, M. I., Moray, N., & Kijowski, B. (1995). Taxonomy of Scheduling Systems as a Basis for the Study of Strategic Behavior. Human Factors, 37(3), 443-472.

Dobson, P., Myles, J., & Jackson, P. (2007). Making the Case for Critical Realism: Examining the Implementation of Automated Performance Management Systems. Information Resources Management Journal, 20(2), 138-152.

Dobson, P. J. (2001). The Philosophy of Critical Realism - An Opportunity for Information Systems Research. Information Systems Frontiers, 3(2), 199-210.

Domschke, W. & Drexl, A. (2007). Einführung in Operations Research. Berlin: Springer.

Dourish, P. B. & Button, G. (1998). On "Technomethodology": foundational relationships between ethnomethodology and system design. Human-Computer Interaction, 13, 395-432.

Dubois, P., Heidenreich, M., La Rosa, M., & Schmidt, G. (1995). New Technologies and Post-Taylorist Regulation Models. The Introduction and Use of Production Planning Systems in French, Italian, and German Enterprises. In W. Littek & T. Charles (Eds.), The New Division of Labour. Berlin: De Gruyter (pp. 287-315).

Dumazeau, C. & Karsenty, L. (2008). Communications Distantes en Situation de Travail: Favoriser l'Etablissement d'un Contexte Mutuellement Partagé. Le Travail Humain, 71(3), 225-252.

Dunckel, H., Volpert, W., Zölch, M., Kreutner, U., Pleiss, C., & Hennes, K. (1992). Kontrastive Aufgabenanalyse im Büro. KABA-Leitfaden: Grundlagen, Manual und Arbeitsblätter. Stuttgart: Teubner.

Dunckel, H., Volpert, W., Zölch, M., Kreutner, U., Pleiss, C., & Hennes, K. (1993). Leitfaden zur kontrastiven Aufgabenanalyse und -gestaltung bei Büro- und Verwaltungstätigkeiten. Das KABA-Verfahren. Zürich: Verlag der Fachvereine.

Dutton, J. M. & Starbuck, W. (1971). Finding Charlie's run-time estimator. In J. M. Dutton & W. Starbuck (Eds.), Computer Simulation of Human Behaviour. Chichester: Wiley.

El Amrani, R., Rowe, F., & Geffroy-Maronnat, B. (2006). The effects of enterprise resource planning implementation strategy on cross-functionality. Information Systems Journal, 16(1), 79-104.

Elbanna, A. R. (2006). The validity of the improvisation argument in the implementation of rigid technology: the case of ERP systems. Journal of Information Technology, 21, 165-175.

Elbanna, A. R. (2008). Strategic systems implementation: diffusion through drift. Journal of Information Technology, 23(2), 89-96.

Elder-Vass, D. (2007). Social Structure and Social Relations. Journal for the Theory of Social Behaviour, 37(4), 463-477.

Elder-Vass, D. (2010). The Causal Power of Social Structures. Cambridge, UK: Cambridge University Press.

Elder-Vass, D. (2012a). Para um construtivismo social realista. Sociologia, Problemas E Práticas, 2012(70).

Elder-Vass, D. (2012b). The reality of social construction. Cambridge: Cambridge University Press.

Ellul, J. (1988). The Global Technological System and the Human Response. Bulletin of Science, Technology & Society, 8(2), 139.

Endsley, M. R., Bolté, B., & Jones, D. G. (2003). Designing for situation awareness: An approach to user-centered design. New York: Taylor & Francis.

Engeström (2004). New forms of learning in co-configuration work. Journal of Workplace Learning, 16(1/2), 11-21.

Erickson & Kellogg (2000). Social Translucence: An Approach to Designing Systems that Support Social Processes. ACM Transactions on Computer-Human Interaction, 7(1), 59-83.

Erlicher, L. & Massone, L. (2005). Human Factors in Manufacturing: New Patterns of Cooperation for Company Governance and the Management of Change. Human Factors and Ergonomics in Manufacturing, 15(4), 403-419.

Ernst, N. A., Jamieson, G. A., & Mylopoulos, J. (2006). Integrating requirements engineering and cognitive work analysis: A case study. In Proceedings of the Fourth Annual Conference on Systems Engineering Research. Los Angeles: INCOSE.

Euerby, A. & Burns, C. M. (2010). Advancing complex sociotechnical systems design using the community of practice concept. In Proceedings of the Human Factors and Ergonomics Society 54th Annual Meeting. San Francisco: HFES.

Farrington-Darby, T., Wilson, J. R., Norris, B. J., & Clarke, T. (2006). A naturalistic study of railway controllers. Ergonomics, 49(12-13), 1370-94.

Feenberg, A. (1999). Questioning Technology. London: Routledge.

Feldman, M. S. & Pentland, B. T. (2003). Reconceptualizing Organizational Routines as a Source of Flexibility and Change. Administrative Science Quarterly, 48(1), 94-121.

Fields, B., Amaldi, P., & Tassi, A. (2005). Representing collaborative work: The airport as common information space. Technology & Work, 7(2), 119-133.

Fine, G. A. (1990). Organizational Time: Temporal Demands and the Experience of Work in Restaurant Kitchens. Social Forces, 69(1), 95-114.

Finney, S. & Corbett, M. (2007). ERP implementation: a compilation and analysis of critical success factors. Business Process Management Journal, 13(3), 329-347.

Flach, J. M. (1995). Situation awareness: Proceed with caution. Human Factors, 37(1), 149-157.

Fleck, J., Webster, J., & Williams, R. (1990). Dynamics of information technology implementation. Futures, 22(6), 618-640.

Fleig, J. & Schneider, R. (1998). The Link to the 'Real World': Information, Knowledge and Experience for Decision Making. In E. Scherer (Ed.), Shop Floor Control: A Systems Perspective. Springer Verlag. (pp. 221-43).

Flemming, S. A. C., Hilliard, A., & Jamieson, G. A. (2008). The Need for Human Factors in the Sustainability Domain. In Human Factors and Ergonomics Society Annual Meeting Proceedings. New York.

Flores, F., Graves, M., Hartfield, B., & Winograd, T. (1988). Computer systems and the design of organizational interaction. ACM Transactions on Information Systems (TOIS), 6(2), 153-172.

Floridi, L. (2011). Harmonising Physis and Techne: The Mediating Role of Philosophy. Philosophy & Technology, 1-3.

Foss, N. & Lorenzen, M. (2009). Towards an understanding of cognitive coordination: Theoretical developments and empirical illustrations. Organization Studies, 30(11), 1201-1226.

Fox, D., Prilleltensky, I., & Austin, S. (2009). Critical psychology: An introduction. Los Angeles: SAGE.

Franklin, U. M. (1990). The Real World of Technology. Toronto: Anansi.

Frese, M. (1988). A theory of control and complexity: implications for software design and integration of computer systems into the work place. In M. Frese, E. Ulich, & W. Dzida (Eds.), Psychological Issues of Human Computer Interaction in the Work Place. Amsterdam: North Holland. (pp. 313-37).

Funken, C. & Schulz-Schaeffer, I. (2008). Digitalisierung der Arbeitswelt: Zur Neuordnung formaler und informeller Prozesse in Unternehmen. Wiesbaden: Verlag für Sozialwissenschaften.

Furniss, D. & Blandford, A. (2006). Understanding emergency medical dispatch in terms of distributed cognition: a case study. Ergonomics, 49(12-13), 1174-203.

Garcia-Retamero, R. & Hoffrage, U. (2006). How causal knowledge simplifies decision-making. Minds and Machines, 16(3), 365-380.

Gasser, R. (2010). Decision Strategy Types and Situation-Contingent Selection Mechanisms: A Review and Some Field Data. In Proceedings of the 54th Meeting of the Human Factors and Ergonomics Society. San Francisco: Human Factors and Ergnomics Society.

Gasser, R., Fischer, K., & Wäfler, T. (2007). Decision support in complex planning environments. In W. Karwowski & S. Trzcielinski (Eds.), International Conference on Human Aspects of Advanced Manufacturing: Agility and Hybrid Automation. Poznan, Poland: IEA Press.

Gasser, R., Fischer, K., & Wäfler, T. (2011). Decision Making in Planning and Scheduling: A Field Study in Planning Behaviour in Manufacturing. In J. Fransoo, T. Wäfler, & J. Wilson (Eds.), Behavioral Operations in Planning and Scheduling. Berlin: Springer (pp. 11-30).

Gasser, R., Gärtner, K., & Wäfler, T. (2008). Kontroll-Kapazität im Unternehmen: Erhebungsinstrumente und Analyseverfahren. Unpublished report. Olten: FHNW / MikS.

Gasser, R., Shepherd, C., & Gärtner, K. (2010). Enterprise resource planning as discursively negotiated technology. Porto, Portugal: 17th EUROMA conference.

Gattiker, T. F. & Goodhue, D. L. (2004). Understanding the local-level costs and benefits of ERP through organizational information processing theory. Information & Management, 41(4), 431-443.

Gergen, K. J. (1992). Toward a Postmodern Psychology. In S. Kvale (Ed.), Psychology and Postmodernism. London: Sage (pp. 17-30).

Gibson, J. J. (1979). The ecological approach to visual perception. Boston: Houghton Mifflin.

Gigerenzer, G. (2000). Adaptive thinking: Rationality in the real world. Oxford University Press, USA.

Gigerenzer, G. (2001). Decision making: Nonrational theories. International Encyclopedia of the Social and Behavioral Sciences, 5, 3304-9.

Gigerenzer, G. (2002). Rationality: Why Social Context Matters. In Adaptive Thinking. Oxford: Oxford University Press. (pp. 201-10).

Gigerenzer, G. & Goldstein, D. G. (1996). Reasoning the fast and frugal way: models of bounded rationality. Psychological Review, 103(4), 650-69.

Gigerenzer, G. & Hug, K. (1992). Domain-specific reasoning: social contracts, cheating, and perspective change. Cognition, 43(2), 127-71.

Gillespie, D. (1992). The mind's we : contextualism in cognitive psychology. Carbondale : Southern Illinois University Press.

Glöckner, A. & Betsch, T. (2008). Modeling option and strategy choices with connectionist networks: Towards an integrative model of automatic and deliberate decision making. Judgment and Decision Making, 3(3), 215-228.

Gosain, S. (2004). Enterprise information systems as objects and carriers of institutional forces: the new iron cage? Journal of the Association for Information Systems, 5(4), 151-182.

Grant, D. & Hall, R. (2005). Power, discourse and enterprise resource planning systems. In T. Lê & M. Short (Eds.), Proceedings of the International Conference on Critical Discourse Analysis. Launceston: University of Tasmania.

Grant, D., Hall, R., Wailes, N., & Wright, C. (2006). The false promise of technological determinism: the case of enterprise resource planning systems. New Technology, Work and Employment, 21(1), 2-15.

Grint, K. & Woolgar, S. (1997). The machine at work: Technology, work, and organization. Cambridge, UK: Blackwell.

Grote, G. (2004). Uncertainty management at the core of system design. Annual Reviews in Control, 28(2), 267-274.

Grote, G., Wäfler, T., Ryser, C., Weik, S., Zölch, M., & Windischer, A. (1999). Wie sich Mensch und Technik sinnvoll ergänzen. Die Analyse automatisierter Produktionssysteme mit KOMPASS. Zürich: vdf Hochschulverlag.

Günter, H. (2007). Collaborative planning in heterarchic supply networks. Thesis, ETH Zürich.

Hacker, W. (1995). Arbeitstätigkeitsanalyse - Analyse und Bewertung psychischer Arbeitsanforderungen. Heidelberg: Asanger.

Hacker, W., Volpert, W., & Cranach, M. V. (1982). Cognitive and motivational aspects of action: XXIInd International Congress of Psychology, Leipzig, July 6-12, 1980. Amsterdam: North-Holland.

Hammond, K. R. (1988). Judgment and decision making in dynamic tasks. Information and Decision Technologies(Amsterdam), 14(1), 3-14.

Hanseth, O. & Monteiro, E. (1997). Inscribing Behavior in Information Infrastructure Standards. Accounting, Management and Information Technology, 7(4), 183-211.

Hanseth, O., Ciborra, C. U., & Braa, K. (2001). The control devolution: ERP and the side effects of globalization. ACM SIGMIS Database, 32(4), 34-46.

Harvey, D. (2004). The principles of dialectics. In W. K. Carroll (Ed.), Critical strategies for social research. Toronto: Canadian Scholars Press (pp. 125-32).

Hayes-Roth, B. & Hayes-Roth, F. (1979). A Cognitive Model of Planning. Cognitive Science, 3, 275-310.

Heath, C., Luff, P., & Knoblauch, H. (2004). Tools, Technologies and Organizational Interaction: The Emergence of 'Workplace Studies'. In D. Grant, C. Hardy, C. Oswick, & L. L. Putnam (Eds.), The Sage handbook of organizational discourse. London: Sage Publications. (pp. 338-58).

Hedström, P. & Swedberg, R. (1998). Social mechanisms: An analytical approach to social theory. New York: Cambridge University Press.

Hendrick, H. W. & Kleiner, B. M. (2002). Macroergonomics: Theory, methods, and applications. Mahwah: Lawrence Erlbaum.

Herrmann, J. W. (2005). A history of decision-making tools for production scheduling. In Multidisciplinary Conference on Scheduling: Theory and Applications. New York.

Herrmann, J. W. (2007). The Legacy of Taylor, Gantt, and Johnson: How to Improve Production Scheduling (TR 2007-26). College Park, ML: The Institute for Systems Research, University of Maryland.

Higgins, P. G. (1999). Job shop scheduling: Hybrid intelligent human-computer paradigm. Thesis, Melbourne: University of Melbourne.

Hildebrandt, M. & Harrison, M. (2004). PaintShop: A Microworld Experiment Investigating Temporal Decisions in a Supervisory Control Task. In Human Factors and Ergonomics Society Annual Meeting Proceedings.

Hill, S. (1981). Competition and Control at Work: A New Industrial Sociology. Cambridge, MA: MIT Press.

Hoc, J. M. (1988). Cognitive psychology of planning. San Diego: Academic Press.

Hoc, J. M. (1993). Some dimensions of a cognitive typology of process control situations. Ergonomics, 36(11), 1445-1455.

Hoc, J. M. (2000). From human-machine interaction to human-machine cooperation. Ergonomics, 43(7), 833-43.

Hoc, J. M. (2001). Towards a cognitive approach to human-machine cooperation in dynamic situations. International Journal of Human-Computer Studies, 54(4), 509-540.

Hoc, J. M. (2006). Planning in dynamic situations: some findings in complex supervisory control. In W. van Wezel, R. Jorna, & A. Meystel (Eds.), Planning in intelligent systems: Aspects, motivations, and methods. New York: Wiley.

Hoc, J. M., Mebarki, N., & Cegarra, J. (2004). L'assistance à l'opérateur humain pour l'ordonnancement dans les ateliers manufacturiers. Le Travail Humain, 67(2).

Hodgkinson, G. P. & Healey, M. P. (2008). Cognition in organizations. Annual Review of Psychology, 59, 387-417.

Hofkirchner, W. (2010). Twenty Questions About a Unified Theory of Information. Litchfield Park: Emergent Publications.

Hogarth, R. M. (2001). Educating intuition . Chicago : University of Chicago Press.

Hogarth, R. M. (2005). Deciding Analytically or Trusting Your Intuition? The Advantages and Disadvantages of Analytic and Intuitive Thought. In T. Betsch & S. Haberstroh (Eds.), The Routines of Decision Making. Mahwah: Lawrence Erlbaum (pp. 39-66).

Hogarth, R. M. (2006). Is Confidence in Decision Related to Feedback? Evidence from Random Samples of Real-World Behavior. In K. Fiedler & P. Juslin (Eds.), Information sampling and adaptive cognition. Cambridge: Cambridge University Press (pp. 456-84).

Hogarth, R. M. (2008). On the Learning of Intuition. In H. Plessner, C. Betsch, & T. Betsch (Eds.), Inituition in Judgment and Decision Making. London: Taylor & Francis (pp. 91-105).

Holland, C. P. & Light, B. (1999). A critical success factors model for ERP implementation. IEEE Software, 16(3), 30-36.

Hollnagel, E. (2007). Decisions about "What" and Decisions about "How". In M. Cook, J. Noyes, & Y. Masakowski (Eds.), Decision making in complex environments. Aldershot: Ashgate (pp. 3-12).

Hollnagel, E. & Woods, D. D. (2005). Joint cognitive systems: Foundations of cognitive systems engineering. Boca Raton: Taylor and Francis.

Holzkamp, K. (1985). Grundlegung der Psychologie. Frankfurt/Main; New York: Campus-Verlag.

Holzkamp, K. (1992). On doing psychology critically. Theory & Psychology, 2(2), 193-204.

Hong, K. K. & Kim, Y. G. (2002). The critical success factors for ERP implementation: an organizational fit perspective. Information & Management, 40, 25-40.

Horn, J. & Masunaga, H. (2006). A Merging Theory of Expertise and Intelligence. In K. A. Ericsson, N. Charness, P. J. Feltovich, & R. R. Hoffman (Eds.), The Oxford Handbook on Expertise and Expert Performance. Cambridge: Cambridge University Press (pp. 587-611).

Howard-Grenville, J. A. (2005). The Persistence of Flexible Organizational Routines: The Role of Agency and Organizational Context. Organization Science, 16(6), 618-636.

Howcroft, D. (2009). Information systems. In M. Alvesson, T. Bridgman, & H. Willmott (Eds.), The Oxford Handbook of Critical Management Studies. Oxford: Oxford University Press (pp. 392-413).

Höök, K., Benyon, D., & Munro, A. J. (2003). Designing information spaces : the social navigation approach. London, New York: Springer.

Hutchby, I. (2001). Technologies, texts and affordances. Sociology, 35(2), 441-456.

Hutchins, E. (1995). Cognition in the Wild. Cambridge: MIT press.

Ifinedo, P., Rapp, B., Ifinedo, A., & Sundberg, K. (2010). Relationships among ERP post-implementation success constructs: An analysis at the organizational level. Comput. Hum. Behav., 26(5), 1136-1148.

Ignatiadis, I. & Nandhakumar, J. (2006). Organizational Work With Enterprise Systems: A Double Agency Perspective. In Proceedings of the 14th European Conference on Information Systems (ECIS'2006). Goteborg: Sweden.

Jackson, S., Wilson, J. R., & MacCarthy, B. L. (2004). A new model of scheduling in manufacturing: Tasks, roles, and monitoring. Human Factors, 46(3), 533-550.

Jamieson, G. A. (2003). Bridging the gap between cognitive work analysis and ecological interface design. In Human Factors and Ergonomics Society Annual Meeting Proceedings. Denver, Colorado.

Jamieson, G. A. & Miller, C. A. (2000). Exploring the "Culture of Procedures". In Proceedings of the 5th International Conference on Human Interaction with Complex Systems. Urbana, Illinois.

Jamieson, G. A. & Vicente, K. J. (2005). Designing Effective Human-Automation-Plant Interfaces: A Control-Theoretic Perspective. Human Factors, 47(1), 12-34.

Jian, G. (2011). Articulating circumstance, identity and practice: toward a discursive framework of organizational changing. Organization, 18(1), 45.

Jones, P. (2009). Breaking away from Capital? Theorising activity in the shadow of Marx. Outlines. Critical Practice Studies, 11(1), 45-58.

Jones, P. E. (2007). Why there is no such thing as "critical discourse analysis". Language & Communication, 27(4), 337-368.

Jorna, R. (2006). Cognition, Planning, and Domains: An Empirical Study into the Planning Processes of Planners. In W. van Wezel, R. Jorna, & A. Meystel (Eds.), Planning in intelligent systems: Aspects, motivations, and methods. New York: Wiley (pp. 101-35).

Kallinikos, J. (1995). The Archi-tecture of the Invisible: Technology is Representation. Organization, 2(1), 117-140.

Kallinikos, J. (2009). On the computational rendition of reality: Artefacts and human agency. Organization, 16(2), 183.

Kallinikos, J., Leonardi, P. M., & Nardi, B. A. (2012). The Challenge of Materiality: Origins, Scope, and Prospects. In P. M. Leonardi, B. A. Nardi, & J. Kallinikos (Eds.), Materiality and Organizing. Oxford: Oxford University Press (pp. 3-24).

Kannabiran, G. & Graves Petersen, M. (2010). Politics at the interface: A Foucauldian power analysis. In NordiCHI2010, October 16-20. Reykjavik, Iceland: ACM (pp. 695-8).

Kaptelinin, V. & Nardi, B. A. (2006). Acting with technology. Cambridge, MA: MIT Press.

Karltun, J. & Berglund, M. (2010). Contextual conditions influencing the scheduler's work at a sawmill. Production Planning & Control, 21(4), 359-374.

Karltun, J., von der Weth, R., & Gasser, R. (2006). Control Capacity Working Paper. Unpublished report. Olten: FHNW / MikS.

Keating, C. B., Fernandez, A. A., Jacobs, D. A., & Kauffmann, P. (2001). A methodology for analysis of complex sociotechnical processes. Business Process Management Journal, 7(1), 33-50.

Keller, R. (2007). Diskurse und Dispositive analysieren. Die Wissenssoziologische Diskursanalyse als Beitrag zu einer wissensanalytischen Profilierung der Diskursforschung. Forum: Qualitative Social Research, 8(2), Art. 19.

Keller, R. (2011). Wissenssoziologische Diskursanalyse. Wiesbaden: VS Verlag für Sozialwissenschaften.

Kellogg, K. C., Orlikowski, W. J., & Yates, J. (2006). Life in the Trading Zone: Structuring Coordination Across Boundaries in Postbureaucratic Organizations. Organization Science, 17(1), 22-44.

De Keyser, V. (1995). Time in ergonomics research. Ergonomics, 38(8), 1639-1660.

De Keyser, V. & Nyssen, A. S. (2001). The Management of Temporal Constraints in Naturalistic Decision Making: The Case of Anesthesia. In E. Salas & G. Klein (Eds.), Linking Expertise and Naturalistic Decision Making. Mahwah: Lawrence Erlbaum.

Kirlik, A. (1995). Requirements for psychological models to support design: toward an ecological task analysis. In J. Flach, P. Hancock, J. Caird, & K. Vicente (Eds.), Global perspectives on the Ecology of Human Manchine Systems. Hillsdale: Lawrence Erlbaum (pp. 68-120).

Kitto, S. & Higgins, V. (2010). Working around ERPs in Technological Universities. Science, Technology & Human Values, 35(1), 29-54.

Kleemann, F. & Matuschek, I. (2008). Informalisierung als Komplement der Informatisierung von Arbeit. In C. Funken & I. Schulz-Schaeffer (Eds.), Digitalisierung der Arbeitswelt. Wiesbaden: Verlag für Sozialwissenschaften (pp. 43-67).

Klein, G. (1998). Sources of Power: How people make decisions. Cambridge: MIT Press.

Klein, G., Orasanu, J., Calderwood, R., & Zsambok, C. E. (1993). Decision Making in Action: Models and Methods. Norwood: Ablex Publishing.

Klein, K. J., Conn, A. B., & Sorra, J. S. (2001). Implementing computerized technology: An organizational analysis. Journal of Applied Psychology, 86(5), 811-824.

Kleinau, T. (2005). Der Rollenwandel im mittleren Management: der Meister als Prozessmanager. Konzeption und Evaluation eines Personal-und Organisationsentwicklungsprojektes zur Förderung der Führungskompetenz in der Automobilindustrie. Thesis. Braunschweig: Technische Universität.

Knoblauch, H. (2005). Wissenssoziologie. Konstanz: UVK Verlagsgesellschaft.

Knoblauch, H. (2013). Grundbegriffe und Aufgaben des kommunikativen Konstruktivismus. In R. Keller, H. Knoblauch, & J. Reichertz (Eds.), Kommunikativer Konstruktivismus: Theoretische und empirische Arbeiten zu einem neuen wissenssoziologischen Ansatz. Wiesbaden: Springer VS. (pp. 25-47).

Knoblauch, H. & Heath, C. (1999). Technologie, Interaktion und Organisation: Die Workplace Studies. Schweizerische Zeitschrift Für Soziologie, 25(2), 163-181.

Knorr Cetina, K. (2009). The Synthetic Situation: Interactionism for a Global World. Symbolic Interaction, 32(1), 61-87.

Koch, C. (2005). Users? What users?-Shaping global corporations and generic users with ERP. UITQ, 51-61.

Koch, C. & Buhl, H. (2001). ERP-Supported Teamworking in Danish Manufacturing? New Technology, Work and Employment, 16(3), 164-177.

Kontogiannis, T. (2010a). Adapting plans in progress in distributed supervisory work: aspects of complexity, coupling, and control. Cognition, Technology & Work, 12(2), 103-118.

Kontogiannis, T. (2010b). A contemporary view of organizational safety: variability and interactions of organizational processes. Cognition, Technology & Work, 12(2), 231-249.

Kvale, S. (1992). Psychology and postmodernism. London: Sage Publications.

Kwahk, K. Y. & Ahn, H. (2010). Moderating effects of localization differences on ERP use: A socio-technical systems perspective. Comput. Hum. Behav., 26(2), 186-198.

Lacau, E. & Mouffe, C. (2001). Hegemony and Socialist Strategy. London: Verso.

Landauer, T. K. (1995). The Trouble with Computers. Cambridge MA: MIT Press.

Latour, B. (1993). We have never been modern. Cambridge MA: Harvard University Press.

Latour, B. (2005). Reassembling the Social. Oxford: Oxford University Press.

Lauche, K. (2008). Neue Formen der Zusammenarbeit. In P. Badke-Schaub, G. Hofinger, & K. Lauche (Eds.), Human Factors. Heidelberg: Springer Verlag.

Law, J. (1991). A Sociology of Monsters: Essays on Power, Technology and Domination. London: Routledge.

Law, J. & Urry, J. (2004). Enacting the social. Economy and Society, 33(3), 390-410.

Lemon, M., Craig, J., & Cook, M. (2010). Looking In or Looking Out? Top-down Change and Operational Capability. Forum: Qualitative Social Research, 11(3), 27.

Leonardi, P. M. (2012). Materiality, Sociomateriality, and Socio-Technical Systems: What Do These Terms Mean? How Are They Different? Do We Need Them? In P. M. Leonardi, B. A. Nardi, & J. Kallinikos (Eds.), Materiality and Organizing. Oxford: Oxford University Press (pp. 3-24).

Leonardi, P. M. (2013). Theoretical foundations for the study of sociomateriality. Information and Organization, 23(2), 59-76.

Leontjew, A. N. (1982). Tätigkeit, Bewusstsein, Persönlichkeit. Köln: Pahl-Rugenstein.

Lewin, K. (1930). Der Übergang von der aristotelischen zur galileischen Denkweise in Biologie und Psychologie. Erkenntnis, 1(1), 421-466.

Lievens, F. & Anseel, F. (2004). Confirmatory factor analysis and invariance of an organizational citizenship behaviour measure across samples in a Dutch-speaking context. Journal of Occupational and Organizational Psychology, 77(3), 299-306.

Light, B. & Wagner, E. (2006). Integration in ERP environments: rhetoric, realities and organisational possibilities. New Technology, Work and Employment, 21(3), 215-228.

Liu, L. & Yu, E. (2004). Designing Information Systems in Social Context: A Goal and Scenario Modelling Approach. Information Systems, 29(2), 187-203.

Locke, J. & Lowe, A. (2007). A Biography: Fabrications in the Life of an ERP Package. Organization Science, 14(6), 793-814.

Lopéz, J. & Potter, G. (2001). After Postmodernism: An Introduction to Critical Realism. London: The Athlone Press.

Luff, P., Hindmarsh, J., & Heath, C. (2000). Workplace Studies: Recovering work practice and informing system design. Cambridge University Press.

Luhmann, N. (2000). Organisation und Entscheidung. Wiesbaden: Opladen.

MacCarthy, B. & Wilson, J. R. (2001). Human Performance in Planning and Scheduling. London: Taylor and Francis.

Malone, T. W., Crowston, K., & Herman, G. A. (2003). Organizing business knowledge : the MIT process handbook. Cambridge, MA: MIT Press.

March, J. G. (1978). Bounded rationality, ambiguity, and the engineering of choice. The Bell Journal of Economics, 587-608.

Marmaras, N. & Nathanael, D. (2005). Cognitive engineering practice: melting theory into reality. Theoretical Issues in Ergonomics Science, 6(2), 109-127.

Martín-Baró, I. (1994). Writings for a liberation psychology. Cambridge, MA: Harvard University Press.

Mata, R., Schooler, L. J., & Rieskamp, J. (2007). The aging decision maker: cognitive aging and the adaptive selection of decision strategies. Psychology and Aging, 22(4), 796-810.

Matern, B. (1984). Psychologische Arbeitsanalyse. Berlin: Springer.

May, C. & Finch, T. (2009). Implementing, embedding, and integrating practices: An outline of Normalization Process Theory. Sociology, 43(3), 535.

McKay (1992). Production Planning and Scheduling: A Model for Manufacturing Decisions Requiring Judgement. In Production

Planning and Scheduling: A Model for Manufacturing Decisions Requiring Judgement. Doctoral Thesis, Waterloo: University of Waterloo.

McKay, K. & Wiers, V. C. (2003). Planning, scheduling and dispatching tasks in production control. Cognition, Technology & Work, 5(2), 82-93.

McKay, K., Buzacott, J. A., & Safayeni, F. R. (1989). The Schedulers Knowledge of Uncertainty: The Missing Link. In J. Browne (Ed.), Knowledge Based Production Management Systems. North-Holland: Elsiever.

McKay, K. N. & Buzacott, J. A. (1995). 'Common sense' realities of planning and scheduling in printed circuit board production. International Journal of Production Research, 33(6), 1587-1603.

McKay, K. N. & Wiers, C. S. (2006). The Organizational Interconnectivity of Planning and Scheduling. In W. van Wezel, R. Jorna, & A. Meystel (Eds.), Planning in intelligent systems: Aspects, motivations, and methods. New York: Wiley (pp. 176-201).

McPherson, R. F. & White, K. P. (2006). A framework for developing intelligent real-time scheduling systems. Human Factors and Ergonomics in Manufacturing, 16(4), 385.

McPherson, R. F. & White Jr, K. P. (1994). Management control and the manufacturing hierarchy: Managing integrated manufacturing organizations. International Journal of Human Factors in Manufacturing, 4(2), 121-144.

Mehnert, J. (2004). Gestaltung und Integration von Arbeitsplanungskompetenzen für hierarchielose Produktionsnetze. Thesis, Chemnitz: TU Chemnitz.

Moon, Y. B. (2007). Enterprise resource planning (ERP): a review of the literature. International Journal of Management and Enterprise Development, 4(3), 235-264.

Morris, R. & Ward, G. (2005). The cognitive psychology of planning. London: Psychology Press.

Moscoso, P., Fransoo, J., Fischer, D., & Wäfler, T. (2011). The Planning Bullwhip: A Complex Dynamic Phenomenon in Hierarchical Systems. In J. Fransoo, T. Wäfler, & J. Wilson (Eds.), Behavioral Operations in Planning and Scheduling. Berlin: Springer (pp. 159-86).

Mutch, A. (2002). Actors and Networks or Agents and Structures: Towards a Realist View of Information Systems. Organization, 9(3), 477-496.

Mutch, A. (2010). Technology, Organization, and Structure--A Morphogenetic Approach. Organization Science, 21(2), 507-520.

Mutch, A. (2013). Sociomateriality — Taking the wrong turning? Information and Organization, 23(1), 28-40.

Møller, C. (2005). ERP II: a conceptual framework for next-generation enterprise systems? Journal of Enterprise Information Management, 18(4), 483-497.

Møller, C., Kræmmergaard, P., Rikhardsson, P., Møller, P., Jensen, T. N., & Due, L. (2004). A comprehensive ERP bibliography-2000-2004 (IFI Working Paper no. 129). Aarhus: Institut for Informationsbehandling, Det Erhvervsøkonomiske Fakultet, Aarhus School of Business.

Naikar, N. (2006). Beyond interface design: Further applications of cognitive work analysis. International Journal of Industrial Ergonomics, 36(5), 423-438.

Nandhakumar, J., Rossi, M., & Talvinen, J. (2005). The dynamics of contextual forces of ERP implementation. The Journal of Strategic Information Systems, 14(2), 221-242.

Nardi, B., Whittaker, S., & Schwarz, H. (2001). NetWORKers and their activity in intensional networks. J. Comput.-Support. Coop. Work, 205-242.

Nardi, B. A. (1996). Studying context: A comparison of Activity Theory, Situated Action Models, and Distributed Cognition. In B. A. Nardi (Ed.), Context and Consciousness. Cambridge: MIT press (pp. 69-102).

Nardi, B. A. & O'Day, V. L. (2000). Information ecologies. Cambridge, MA: MIT Press.

Nauta, A. & Sanders, K. (2001). Causes and consequences of perceived goal differences between departments within manufacturing organizations. Journal of Occupational and Organizational Psychology, 74, 321-342.

Noble, D. F. (1984). Forces of production: A social history of industrial automation. Oxford: Oxford University Press.

Norros, L. (1995). An Orientation-Based Approach to Expertise. In J.-M. Hoc, P. C. Cacciabue, & E. Hollnagel (Eds.), Expertise and Technology. Hillsdale NJ: Lawrence Erlbaum (pp. 141-164).

Nyssen, A. S. & Javaux, D. (1996). Analysis of synthronization constraints and associated errors in collective work environments. Ergonomics, 39(10), 1249-1264.

Oborski, P. (2004). Man-machine interactions in advanced manufacturing systems. The International Journal of Advanced Manufacturing Technology, 23(3), 227-232.

Oesterreich, R. (1991). VERA, Version 2: Arbeitsanalyseverfahren zur Ermittlung von Planungs- und Denkanforderungen im Rahmen der RHIA-Anwendung. Berlin: TU Berlin.

Orlikowski (2002). Knowing in practice: Enacting a collective capability in distributed organizing. Organization Science, 13(3), 249-273.

Orlikowski, W. & Barley, S. (2001). Technology and institutions: What can research on information technology and research on organizations learn from each other? MIS Quarterly, 25(2), 145-65.

Orlikowski, W. J. (1992). The duality of technology: Rethinking the concept of technology in organizations. Organization Science, 3(3), 398-427.

Orlikowski, W. J. (2000). Using technology and constituting structures: A practice lens for studying technology in organizations. Organization Science, 11(4), 404-428.

Orlikowski, W. J. (2005). Material works: Exploring the situated entanglement of technological performativity and human agency. Scandinavian Journal of Information Systems, 17(1), 183-186.

Orlikowski, W. J. (2007). Sociomaterial practices: Exploring technology at work. Organization Studies, 28(9), 1435-1448.

Orlikowski, W. J. & Barley, S. R. (2001). Technology and Institutions: What Can Research on Information Technology and Research on Organizations Learn from Each Other? MIS Quarterly, 25(2), 145-165.

Orlikowski, W. J. & Gash, D. C. (1994). Technological Frames: Making Sense of Information Technology in Organizations. ACM Transactions on Information Systems, 12(2), 174-207.

Orlikowski, W. J. & Scott, S. V. (2008). Sociomateriality: Challenging the Separation of Technology, Work and Organization. London: London School of Economics. Departement of Management Working Paper Series.

Paay, J., Sterling, L., Vetere, F., Howard, S., & Boettcher, A. (2009). Engineering the social: The role of shared artifacts. International Journal of Human-Computer Studies, 67(5), 437-454.

Parasuraman, R. & Wickens, C. D. (2008). Humans: Still vital after all these years of automation. Human Factors: The Journal of the Human Factors and Ergonomics Society, 50(3), 511.

Parker, I. (2005a). Lacanian Discourse Analysis in Psychology: Seven Theroetical Elements. Theory & Psychology, 15(2), 163-182.

Parker, I. (2005b). Qualitative psychology: Introducing radical research. Maidenhead, UK: Open University Press.

Parker, I. (2007). Revolution in psychology: Alienation to emancipation. London: Pluto Press.

Parker, I. & Shotter, J. (1990). Deconstructing social psychology. London: Routledge.

Parunak, H. V. D. (1991). Characterizing the manufacturing scheduling problem. Journal of Manufacturing Systems, 10(3), 241-259.

Payne, D. G. & Blackwell, J. M. (1997). Toward a Valid View of Human Factors Research: Response to Vicente (1997). Human Factors, 39(2), 329-331.

Payne, J. W., Bettman, J. R., & Johnson, E. J. (1993). The adaptive decision maker. Cambridge: Cambridge University Press.

Pearce, F. & Woodiwiss, T. (2001). Reading Foucault as a Realist. In J. Lopéz & G. Potter (Eds.), After Postmodernism: An Introduction to Critical Realism. London: Athlone Press (pp. 51-62).

Pentland, B. T. & Feldman, M. S. (2005). Organizational routines as a unit of analysis. Industrial and Corporate Change, 14(5), 793-815.

Pettersen, K. A., McDonald, N., & Engen, O. A. (2010). Rethinking the role of social theory in socio-technical analysis: a critical realist approach to aircraft maintenance. Cognition, Technology & Work, (12), 181-191.

Pettersson, G. (2007). Time and Design in Decision Making Environments. In M. Cook, J. Noyes, & Y. Masakowski (Eds.), Decision making in complex environments. Aldershot: Ashgate (pp. 33-42).

Pinch, T. (2008). Technology and institutions: living in a material world. Theory & Society, 37, 461-483.

Pollock, N. & Williams, R. (2009). The sociology of a market analysis tool: How industry analysts sort vendors and organize markets. Information and Organization, 19(2), 129-151.

Porter, J. K., Jarvis, P., Little, D., Lookmann, J., Hannen, C., & Schotten, M. (1996). Production planning and control system developments in Germany. International Journal of Operations & Production Management, 16(1), 27-39.

Potter, J. (1996). Representing reality : discourse, rhetoric and social construction. London: Sage.

Radder, H. (1992). Normative Reflexions on Constructivist Approaches to Science and Technology. Social Studies of Science, 22, 141-173.

Rammert, W. (2003). Technik in Aktion: Verteiltes Handeln in soziotechnischen Konstellationen. In T. Christaller & J. Wehner (Eds.), Autonome Maschinen. Wiesbaden: Westdeutscher Verlag (pp. 289-315).

Rammert, W. (2007). Technik - Handeln - Wissen: Zu einer pragmatischen Technik- und Sozialtheorie. Wiesbaden: VS Verlag für Sozialwissenschaften.

Rasmussen, J. (1986). Information Processing and Human-Machine Interaction: An Approach to Cognitive Engineering. Amsterdam: North-Holland.

Rasmussen, J. (1997). Risk Management in a Dynamic Society: A Modelling Problem. Safety Science, 27(2/3), 183-213.

Rasmussen, J., Pejtersen, A. M., & Goodstein, L. P. (1994). Cognitive Systems Engineering. New York: John Wiley & Sons.

Rasmussen, J., Pejtersen, A. M., & Schmidt, K. (1990). Taxonomy for Cognitive Work Analysis. Roskilde: Risø National Laboratory.

Rawls, A. W. (2008). Harold Garfinkel, ethnomethodology and workplace studies. Organization Studies, 29(5), 701.

Reed, M. I. (1997). In Praise of Duality and Dualism: Rethinking Agency and Structure in Organizational Analysis. Organization Studies, 18(1), 1-42.

Reed, M. I. (2009). Critical Realism in Critical Management Studies. In M. Alvesson, T. Bridgman, & H. Willmott (Eds.), The Oxford Handbook of Critical Management Studies (pp. 52-75).

Resch, M. (1988). Die Handlungsregulation geistiger Arbeit. Huber.

Riedel, R., Fransoo, J., Wiers, V., Fischer, K., Cegarra, J., & Jentsch, D. (2011). Building Decision Support Systems for Acceptance. In J. Fransoo, T. Wäfler, & J. Wilson (Eds.), Behavioral Operations in Planning and Scheduling. Berlin: Springer (pp. 231-95).

Riegel, K. F. (1978). Psychology mon amour. Boston: Houghton Mifflin Company.

Rieskamp, J. & Otto, P. E. (2006). SSL: A theory of how people learn to select strategies. Journal of Experimental Psychology. General, 135(2), 207-36.

Rimann, M. & Udris, I. (1997). Subjektive Arbeitsanalyse: Der Fragebogen SALSA. In O. Strohm & E. Ulich (Eds.), Unternehmen arbeitspsychologisch bewerten. Ein Mehr-Ebenen-Ansatz unter besonderer Berücksichtigung von Mensch, Technik und Organisation. Zürich: vdf Hochschulverlag. (pp. 281-98).

Rittenbruch, M., Viller, S., & Mansfield, T. (2007). Announcing activity: Design and evaluation of an intentionally enriched awareness service. Human-Computer Interaction, 22(1), 137-171.

Rognin, L., Salembier, P., & Zouinar, M. (2000). Cooperation, reliability of socio-technical systems and allocation of function. International Journal of Human-Computer Studies, 52(2), 357-379.

Rose, J. & Jones, M. (2005). The double dance of agency: A socio-theoretic account of how machines and humans interact. Systems, Signs & Actions, 1(1), 19-37.

Rowlands, B. (2009). A social actor understanding of the institutional structures at play in information systems development. Information Technology & People, 22(1), 51-62.

Sanderson, P. M. (1989). The Human Planning and Scheduling role in Advanced Manufacturing Systems: An Emerging Human Factors Domain. Human Factors, 31(6), 635-666.

Sanderson, P. M. (1991). Towards the model human scheduler. International Journal of Human Factors in Manufacturing, 1(3).

Sawchuk, P. H., Duarte, N., & Elhammoumi, M. (2006). Critical perspectives on activity: Explorations across education, work, and everyday life. New York: Cambridge University Press.

Schlichter, B. R. & Kraemmergaard, P. (2010). A comprehensive literature review of the ERP research field over a decade. Journal of Enterprise Information Management, 23(4), 486-520.

Schlick, C. M., Bruder, R., & Luczak, H. (2010). Arbeitswissenschaft. Berlin: Springer Verlag.

Schneider, H. M., Buzacott, J. A., & Rücker, T. (2005). Operative Produktionsplanung und -steuerung. München: Oldenbourg.

Schönsleben, P. (2004). Integrales Logistikmanagement: Planung und Steuerung der umfassenden Supply Chain. Berlin: Springer.

Schraube, E. (2009). Technology as materialized action and its ambivalences. Theory & Psychology, 19(2), 296-312.

Schuh, G. (2006). Produktionsplanung und -steuerung. Berlin: Springer.

Scott, W. R. (1987). Organizations: rational, natural, and open systems. Englewood Cliffs: Prentice-Hall.

Shalin, V. L. (2005). The roles of humans and computers in distributed planning for dynamic domains. Cognition, Technology & Work, 7(3), 198-211.

Shalin, V. L. & McCraw, P. M. (2003). Representations for distributed planning. In E. Hollnagel (Ed.), Handbook of cognitive task design. Mahwah: Lawrence Erlbaum (pp. 701-25).

Sharples, M., Jeffery, N., Du Boulay, J. B. H., Teather, D., Teather, B., & Du Boulay, G. H. (2002). Socio-cognitive engineering: a methodology for the design of human-centred technology. European Journal of Operational Research, 136(2), 310-323.

Shehab, E. M., Sharp, M. W., Supramaniam, L., & Spedding, T. A. (2004). Enterprise resource planning: An integrative review. Business Process Management Journal, 10(4), 359-386.

Shepherd, C. (2006). Constructing enterprise resource planning: A thoroughgoing interpretivist perspective on technological change. Journal of Occupational and Organizational Psychology, 79(3), 357-376.

Shepherd, C., Clegg, C., & Stride, C. (2009). Opening the black box: A multi-method analysis of an enterprise resource planning implementation. Journal of Information Technology, 24(1), 81-102.

Simon, H. A. (1956). Rational choice and the structure of the environment. Psychological Review, 63(2), 129-38.

Simon, H. A. & Newell, A. (1958). Heuristic problem solving: The next advance in operations research. Operations Research, 1-10.

Smith, J. H., Cohen, W. J., Conway, F. T., Carayon, P., Derjani Bayeh, A., & Smith, M. J. (2002). Community Ergonomics. In H. W. Hendrick & B. M. Kleiner (Eds.), Macroergonomics: Theory, methods, and applications. Mahwah: Lawrence Erlbaum (pp. 289-309).

Smith, P. J., Bennett, K. B., & Stone, B. R. (2006). Representation Aiding to Support Performance on Problem-Solving Tasks. Reviews of Human Factors and Ergonomics, 2(1), 74-108.

Soja, P. (2006). Success factors in ERP systems implementations. Journal of Enterprise Information Management, 19(4), 418-433.

Somers, T. M. & Nelson, K. G. (2004). A taxonomy of players and activities across the ERP project life cycle. Information & Management, 41, 257-278.

Sougné, J., Nyssen, A. S., & de Keyser, V. (1993). Temporal reasoning and reasoning theories: A case study in anesthesiology. Psychological Belgica, 33, 311-328.

Star, S. L. & Strauss, A. (1999). Layers of silence, arenas of voice: The ecology of visible and invisible work. Computer Supported Cooperative Work (CSCW), 8(1), 9-30.

Sterman, J. & Sweeney, L. B. (2005). Managing complex dynamic systems: Challenge and opportunity for naturalistic decision-making theory. How Professionals Make Decisions, 57-90.

Strauch, B. (2010). Can Cultural Differences Lead to Accidents? Team Cultural Differences and Sociotechnical System Operations. Human Factors, 52(2), 246-263.

Streatfield, P. (2001). The paradox of control in organizations. London: Routledge.

Strohm, O. (1996). Produktionsplanung und-steuerung im Industrieunternehmen aus arbeitspsychologischer Sicht. Zürich: vdf Hochschulverlag.

Strohm, O. & Ulich, E. (1997). Unternehmen arbeitspsychologisch bewerten: Ein Mehr-Ebenen-Ansatz unter besonderer Berucksichtigung von Mensch, Technik und Organisation. Zürich: vdf Hochschulverlag.

Suchman, L. (1994). Do categories have politics? Computer Supported Cooperative Work (CSCW), 2(3), 177-190.

Suchman, L. (1997). Centers of Coordination: A Case and some Themes. In L. B. Resnick, R. Säljö, C. Pontecorvo, & B. Burge (Eds.), Discourse, Tools, and Reasoning. Berlin: Springer Verlag. (pp. 41-62).

Suchman, L., Blomberg, J., Orr, J. E., & Trigg, R. (1999). Reconstructing Technologies as Social Practice. American Behavioral Scientist, 43(3), 392-408.

Suchman, L. A. (1987). Plans and situated actions: The problem of human-computer communication. New York: Cambridge University Press.

Sumner, M. R. (2009). How alignment strategies influence ERP project success. Enterprise Information Systems, 3(4), 425-448.

Tan, D. S., Gergle, D., Mandryk, R., Inkpen, K., Kellar, M., Hawkey, K., et al. (2008). Using job-shop scheduling tasks for evaluating collocated collaboration. Personal and Ubiquitous Computing, 12(3), 255-267.

Teo, T. (1998). Prolegomenon to a Contemporary Psychology of Liberation. Theory & Psychology, 8(4), 527-547.

Teo, T. (2008). From speculation to epistemological violence in psychology: A critical-hermeneutic reconstruction. Theory & Psychology, 18(1), 47.

Teo, T. (2009). Philosophical Concerns in Critical Psychology. In D. Fox, I. Prilleltensky, & S. Austin (Eds.), Critical Psychology. London: Sage Publications (pp. 36-53).

Theureau, J. (1992). Le Cours d'Action: Analyses Sémio-Logique. Bern: Peter Lang.

Theureau, J. (2000). Anthropologie cognitive et analyse des competences. In J. M. Barbier et al. (Eds.), L'analyse de la singularité de l'action. Paris: PUF (pp. 171-211).

Tolman, C. W. (1994). Psychology, society, and subjectivity : an introduction to German critical psychology. London: Routledge.

Tolman, C. W. (2009). Holzkamp's Critical Psychology as a Science from the Standpoint of the Human Subject. Theory & Psychology, 19(2), 149-160.

Tolman, C. W. & Maiers, W. (1991). Critical psychology: contributions to an historical science of the subject. Cambridge: Cambridge Univ. Press.

Trentesaux, D., Moray, N., & Tahon, C. (1998). Integration of the human operator into responsive discrete production management systems. European Journal of Operational Research, 109(2), 342-361.

Trist, E. & Murray, H. (1997). The Social Engagement of Social Science: A Tavistock Anthology. Philadelphia: University of Pennsylvania Press.

Trist, E. L. & Bamforth, K. W. (1951). Some Social and Psychological Consequences of the Longwall Method of Coal-Getting. Human Relations, 4(1), 3.

Udo, G. G. & Ebiefung, A. A. (1999). Human Factors Affecting the Success of Advanced Manufacturing Systems. Computers & Industrial Engineering, 37, 297-300.

Ulich, E. (2005). Arbeitspsychologie. Zürich: vdf Hochschulverlag.

Umble, Haft, & Umble (2003). Enterprise resource planning: Implementation procedures and critical success factors. European Journal of Operational Research, 146, 241-257.

Upton, C. & Doherty, G. (2008). Extending Ecological Interface Design principles: A manufacturing case study. International Journal of Human-Computer Studies, 66(4), 271-286.

Usher, J. M. & Kaber, D. B. (2000). Establishing information requirements for supervisory controllers in a flexible manufacturing system using GTA. Human Factors and Ergonomics in Manufacturing, 10(4), 431-452.

Valorinta, M. (2009). Information technology and mindfulness in organizations. Industrial and Corporate Change, 18(5), 963-997.

Valsiner, J. (2009). Integrating psychology within the globalizing world: A requiem to the post-modernist experiment with Wissenschaft. Integrative Psychological and Behavioral Science, 43(1), 1-21.

Valsiner, J. & van der Veer, R. (2000). The Social Mind. Cambridge, UK: Cambridge University Press.

Vandenberghe, F. (2002). Reconstructing humans: a humanist critique of actant-network theory. Theory, Culture & Society, 19(5-6), 51-67.

Vandierendonck, A. & De Vooght, G. (1998). Mental models and working memory in temporal reasoning and spatial reasoning. In V. de Keyser, G. d'Ydevalle, & A. Vandierendonck (Eds.), Time and the dynamic control of behavior . Göttingen: Hogrefe & Huber (pp. 383-402).

van Dijk, T. A. (1995). Discourse analysis as ideology analysis. In C. Schäffner & A. Wenden (Eds.), Language and peace. Aldershot: Dartmouth (pp. 17-33).

van Dijk, T. A. (1997). Cognitive context models and discourse. In M. I. Stamenov (Ed.), Language Structure, Discourse and the Access to Consciousness. Amsterdam: John Benjamins (pp. 189-226).

van Dijk, T. A. (2001). Critical discourse analysis. In D. Tannen, D. Schiffrin, & H. Hamilton (Eds.), The Handbook of discourse analysis. Oxford: Blackwell (pp. 352-71).
van Dijk, T. A. (2002). Critical Discourse Studies: A Sociocognitive Approach (shortened version). In R. Wodak & M. Meyer (Eds.), Methods for Critical Discourse Analysis. London: Sage (pp. 62-86).
van Dijk, T. A. (2008a). Discourse and context: A sociocognitive approach. Cambridge, UK: Cambridge University Press.
van Dijk, T. A. (2008b). Discourse and power. New York: Palgrave Macmillan.
van Eijnatten, F. M. & van der Zwaan, A. H. (1998). The Dutch IOR approach to organizational design: An alternative to business process re-engineering? Human Relations, 51(3), 289-318.
van Fenema, P. C. (2005a). Collaborative elasticity and breakdowns in high reliability organizations: contributions from distributed cognition and collective mind theory. Cognition, Technology & Work, 7(2), 134-140.
van Fenema, P. C. (2005b). Reconsidering Structure and Agency for Explaining Routine Coordination. In Second conference on organizational routines. Nice, F.
van Leeuwen, T. (2008). Discourse and practice: New tools for critical discourse analysis. Oxford: Oxford University Press.
van Stijn, E. (2006). Miracle or Mirage? An exploration of the pervasive ERP system phenomenon informed by the notion of conflicting memories. Doctoral thesis, Enschede: Eveline van Stijn.
van Wezel, W. & Jorna, R. (2009). Cognition, tasks and planning: supporting the planning of shunting operations at the Netherlands Railways. Cognition, Technology & Work, 11(2), 165-176.
van Wezel, W. & Jorna, R. J. (2001). Paradoxes in planning. Engineering Applications of Artificial Intelligence, 14(3), 269-286.
van Wezel, W., Cegarra, J., & Hoc, J. -M. (2011). Allocating Functions to Human and Algorithm in Scheduling. In J. Fransoo, T. Wäfler, & J. Wilson (Eds.), Behavioral Operations in Planning and Scheduling. Berlin: Springer (pp. 339-70).
Vester, F. (2005). Sensitivitätsmodell Prof. Vester. München: Frederic Vester GmbH.
Vicente, K. J. (1997). Heeding the Legacy of Meister, Brunswik, & Gibson: Toward a Broader View of Human Factors Research. Human Factors, 39(2).
Vicente, K. J. (1999). Cognitive Work Analysis. Mahwah: Lawrence Erlbaum.

Vicente, K. J. (2000). Toward Jeffersonian research programmes in ergonomics science. Theoretical Issues in Ergonomics Science, 1(2), 93-112.
Vicente, K. J. (2002). Ecological interface design: Progress and challenges. Human Factors, 44(1), 62-78.
Vicente, K. J. (2008). Human factors engineering that makes a difference: leveraging a science of societal change. Theoretical Issues in Ergonomics Science, 9(1), 1-24.
Vicente, K. J. & Burns, C. M. (1996). Evidence for direct perception from cognition in the wild. Ecological Psychology, 8(3), 269-280.
Vicente, K. J. & Rasmussen, J. (1990). The ecology of human-machine systems II: Mediating "direct perception" in complex work domains. Ecological Psychology, 2(3), 207-249.
Viller, S. & Sommerville, I. (1999). Coherence: an approach to representing ethnographic analyses in systems design. Human-Computer Interaction, 14(1), 9-41.
Volkoff, O., Strong, D. M., & Elmes, M. B. (2007). Technological embeddedness and organizational change. Organization Science, 18(5), 832-848.
Vollmann, T. E., Berry, W. L., Whybark, D. C., & Jacobs, F. R. (2005). Manufacturing planning and control systems for supply chain management. Boston: McGraw-Hill.
Volpert, W. (1987). Psychische Regulation von Arbeitstätigkeiten. In U. Kleinbeck & J. Rutenfranz (Eds.), Arbeitspsychologie. Göttingen: Hogrefe (pp. 1-42).
von der Weth, R. (2001). Management der Komplexität. Bern: Hans Huber.
Wagner, E. L., Newell, S., & Piccoli, G. (2010). Understanding project survival in an ES environment: a sociomaterial practice perspective. Journal of the Association for Information Systems, 11(5), 276-297.
Wallace, J. R., Scott, S. D., Stutz, T., Enns, T., & Inkpen, K. (2009). Investigating teamwork and taskwork in single-and multi-display groupware systems. Personal and Ubiquitous Computing, 1-13.
Wäfler, T. (2001). Planning and Scheduling in Secondary Work Systems. In B. L. MacCarthy & J. R. Wilson (Eds.), Human Performance in Planning and Scheduling. London: Taylor & Francis (pp. 411-47).
Wäfler, T. (2002). Verteilt koordinierte Autonomie und Kontrolle. Thesis, Zürich: Universität Zürich.
Wäfler, T., Karltun, J., Gärtner, K., Gasser, R., & Bruch, J. (2008). Control Capability in Planning, Scheduling, and Control - A

Design Model. In Human and Organizational Factors in Planning and Scheduling Conference. Lausanne.

Wäfler, T., von der Weth, R., Karltun, J., Starker, U., Gärtner, K., Gasser, R., et al. (2011). Human Control Capabilities. In J. Fransoo, T. Wäfler, & J. Wilson (Eds.), Behavioral Operations in Planning and Scheduling. Berlin: Springer (pp. 199-230).

Wäfler, T., Windischer, A., Ryser, C., Weik, S., & Grote, G. (1999). Wie sich Mensch und Technik sinnvoll ergänzen. Zürich: vdf Hochschulverlag.

Webster, J. (1991a). Advanced manufacturing technologies: Work organisation and social relations crystalised. In J. Law (Ed.), A Sociology of monsters: essays on power, technology, and domination. Routledge (pp. 192-222).

Webster, J. (1991b). Power, technology and flexibility in organizations. In J. Law (Ed.), Advanced manufacturing technologies: Work organisation and social relations crystallised. London: Routledge (pp. 192-222).

Wehner, T., Clases, C., & Bachmann, R. (2000). Co-operation at work: a process-oriented perspective on joint activity in inter-organizational relations. Ergonomics, 43(7), 983-97.

Weick, K. E. & Sutcliffe, K. M. (2001). Managing the unexpected: Assuring high performance in an age of complexity. Jossey-Bass San Francisco.

Wenger, E. (1999). Communities of practice: Learning, meaning, and identity. Cambridge: Cambridge University Press.

Wenger, E., White, N., & Smith, J. D. (2009). Digital habitats: Stewarding technology for communities. Portland: CPsquare.

Wickens, C. D. & Hollands, J. G. (2000). Engineering psychology and human performance. Upper Saddle River: Prentice Hall.

Wiers, V. C. (1997). Human Computer Interaction in Production Scheduling. Thesis, Eindhoven: Technische Universiteit.

Wiers, V. C. S. (1996). A quantitative field study of the decision behaviour of four shopfloor schedulers. Production Planning & Control, 7(4), 383-392.

Willmott, H. (2005). Theorizing contemporary control: some post-structuralist responses to some critical realist questions. Organization, 12(5), 747-780.

Wilson, J. R., Jackson, S., & Nichols, S. (2003). Cognitive work investigation and design in practice: the influence of social context and social work artefacts. In E. Hollnagel (Ed.), Handbook of Cognitive Task Design. Mahwah: Lawrence Erlbaum (pp. 83-98).

Windischer, A. (2003). Kooperatives Planen: Theoretische Herleitung und empirische Überprüfung von Merkmalen und Einflussgrössen

kooperativer Planungstätigkeiten in der abteilungsübergreifenden Bedarfsplanung. Thesis, Zürich: University of Zürich.
Windischer, A., Grote, G., Mathier, F., Meunier Martins, S., & Glardon, R. (2009). Characteristics and organizational constraints of collaborative planning. Cognition, Technology & Work, 11(2), 87-101.
Winner, L. (1980). Do Artifacts Have Politics? Daedalus, 109(1), 121-136.
Winograd, T. & Flores, F. (1986). Understanding computers and cognition: A new foundation for design. Norwood, NJ: Ablex Publishing Corporation.
Winter, R. (2009). Ein Plädoyer für kritische Perspektiven in der qualitativen Forschung. Forum: Qualitative Social Research, 12(1), Art. 7.
Wodak, R. (2006). Mediation between discourse and society: assessing cognitive approaches in CDA. Discourse Studies, 8(1), 179-190.
Wolf, L. D., Potter, P., Sledge, J. A., Boxerman, S. B., Grayson, D., & Evanoff, B. (2006). Describing nurses' work: combining quantitative and qualitative analysis. Human Factors, 48(1), 5-14.
Woods, D. D. & Hollnagel, E. (2006). Joint cognitive systems: Patterns in cognitive systems engineering. Boca Raton: CRC/Taylor & Francis.
Yates, J. F. & Tschirhart, M. D. (2006). Decision-Making Expertise. In K. A. Ericsson, N. Charness, P. J. Feltovich, & R. R. Hoffman (Eds.), The Oxford Handbook on Expertise and Expert Performance. Cambridge: Cambridge University Press (pp. 421-38).
Yu, E. (1995). Modelling strategic relationships for process reengineering. In DKBS-TR-94-6.pdf. Doctoral thesis. Toronto: University of Toronto.
Yu, E. (1997). Towards modelling and reasoning support for early-phase requirements engineering. In Proceedings of the 3rd IEEE International Symposium on Requirements Engineering (RE'97). Annapolis, MD.
Yu, E. (2009). Social Modeling and i*. In D. Hutchison, T. Kanade, & J. Kittler (Eds.), Conceptual Modeling: Foundations and Applications. Berlin: Springer (pp. 99-121).
Zahavi, D. (2005). Subjectivity and Selfhood. Cambridge, MA: MIT Press.
Zimolong, B. (2006). Einführung. In N. Birbaumer, D. Frey, J. Kuhl, B. Zimolong, & U. Konradt (Eds.), Ingenieurpsychologie (Enzyklopädie der Psychologie ed.). Göttingen: Hogrefe (pp. 3-31).

Zölch, M. (2001). Zeitliche Koordination in der Produktion: Aktivitäten der Handlungsverschränkung. Bern: Hans Huber.

Zsambok, C. E. & Klein, G. A. (1997). Naturalistic decision making. Mahwah: Lawrence Erlbaum.

Zuboff, S. (1988). In the Age of the Smart Machine. New York: Basic Books.

www.ingramcontent.com/pod-product-compliance
Lightning Source LLC
Chambersburg PA
CBHW060828170526
45158CB00001B/112